John F. W. Silk

A Manual of Nitrous Oxide Anaesthesia

for the use of students and general practitioners

John F. W. Silk

A Manual of Nitrous Oxide Anaesthesia
for the use of students and general practitioners

ISBN/EAN: 9783337340100

Printed in Europe, USA, Canada, Australia, Japan

Cover: Foto ©berggeist007 / pixelio.de

More available books at **www.hansebooks.com**

A MANUAL

OF

NITROUS OXIDE ANÆSTHESIA

FOR THE USE OF STUDENTS AND GENERAL
PRACTITIONERS

BY

J. FREDᴷ. W. SILK, M.D. (Lond.), &c.

ANÆSTHETIST TO THE GREAT NORTHERN CENTRAL HOSPITAL, AND TO THE NATIONAL
DENTAL HOSPITAL;
PHYSICIAN TO THE ST. PANCRAS AND NORTHERN DISPENSARY.

"Will you laugh me asleep, for I am very heavy?"
"TEMPEST," Act ii, Scene 1.

LONDON
J. & A. CHURCHILL
11, NEW BURLINGTON STREET
1888

To

DR. URBAN PRITCHARD,

PROFESSOR OF AURAL SURGERY AT KING'S COLLEGE.

WHO HAS ADDED TO THE MANY ACTS OF KINDNESS FOR WHICH I AM
ALREADY HIS DEBTOR, BY ALLOWING ME TO DEDICATE TO
HIM THIS HUMBLE BUT SINCERE TRIBUTE OF
RESPECT AND ADMIRATION.

PREFACE.

In those Medical Schools in which it is considered advisable to give instruction in the administration of Anæsthetics, the attention of the student is mostly directed towards Ether, Chloroform, and their allies; in ordinary text-books on Surgery a paragraph of fifteen or twenty lines suffices to discuss the whole subject of Nitrous Oxide, and in more ambitious works on "Anæsthetics" a chapter of as many pages is considered ample. Thus it can hardly be said that the student or practitioner is overburdened with information.

At the same time I venture to think, that a fuller and more detailed account as to the methods employed, will not be altogether unacceptable to a large class of readers, and if this little work serves no other purpose, than that of directing the attention of abler writers than myself to an important branch of our profession, it will not have been written altogether in vain.

In the first three chapters, in which I have attempted to give an account of the properties and mode of action of the gas, I have presupposed a certain amount of technical knowledge on the part of my readers, but with this exception I have entered fully, and it may be thought tediously, into

details, in the hope that by so doing I might possibly enhance
the value of a work, the shortcomings of which in other
directions are sufficiently apparent.

In my endeavour to obtain the most recent and most
reliable information upon the more strictly scientific portion
of my subject, I have quoted freely from the researches and
papers of others, and where possible I have always acknow-
ledged my obligation; if by accident I have neglected to do
so in any instance, I trust the offence (inasmuch as it is un-
intentional) will be pardoned. To Dr. Dudley Buxton I am
particularly indebted, not only for the free use I have made
of his very admirable papers on this subject, but also for his
courtesy in permitting me to use the pulse tracings on page
16. Dr. George Johnson too, has had the kindness to revise
the portion that refers more especially to his own observations upon the mode of action of the gas.

The more practical chapters of the work, describe the
apparatus and methods of my own ordinary everyday procedure.

My thanks are also due to Messrs. Ash & Sons, Messrs,
Barth, and Messrs. Meyer & Meltzer for the loan of the woodcuts
depicted; and last, but by no means least, I must acknowledge
the valuable aid afforded me by my friend Mr. F. B. Leeder
in passing these sheets through the press.

Finally, I submit this little book to the judgment of my
professional brethren, with all due appreciation of its manifold
imperfections and demerits, but in the hope that it may at
least meet with their courteous and kindly consideration.

National Dental Hospital,
    149, Great Portland Street, W.

CONTENTS.

CHAPTER I.

PROPERTIES—PREPARATION—HISTORY.

CHAPTER II.

PHYSIOLOGY AND PATHOLOGY.

CHAPTER III.

MODE OF ACTION.

CHAPTER IV.

APPARATUS.

LIST OF ILLUSTRATIONS.

A MANUAL

OF

NITROUS OXIDE ANÆSTHESIA.

CHAPTER I.

PROPERTIES—PREPARATION—HISTORY.

A KNOWLEDGE of the physical and chemical properties of any therapeutic agent is generally considered necessary, in order to be able to administer it in the proper and most advantageous form, and an acquaintance with the physiological and pathological changes induced by its use is equally essential, before we can understand, or fully appreciate the clinical phenomena observed during its administration; much more is such knowledge requisite if we propose to explain, or even theorise upon its mode of action. We must therefore, before proceeding to the more practical portions of our subject, ascertain what is known concerning nitrous oxide from these standpoints.

CHEMISTRY.

Nitrous oxide, nitrogen monoxide, protoxide of nitrogen, or laughing gas (French *oxyde azoteux* or *protoxyde d'azote*,

B

German *stickstoffoxydul* or *stickoxydul*), is one of the many compounds of the gases nitrogen and oxygen, and is represented by the formula N_2O.

It is a colourless gas, feebly refrangible, with a faintly sweetish taste and smell, and is heavier than air (S. G. 1·527 *air* = 1).

It is somewhat soluble in cold water, one volume of which at 0° C. (32° F.) absorbs 1·305 volumes of the gas, and as the temperature of the water is raised, the solubility diminishes, until at 24° C. (75° F.) one volume of water only dissolves ·608 volume of the gas, and it is less soluble still in brine or mercury; in alcohol, ether, or oil, it is rather more soluble than in cold water. It is not in itself inflammable, but appears to support combustion in so far as it will re-kindle a glowing chip of wood, or will intensify the brilliancy of any flame thrust into it, but this is owing to the ease with which comparatively slight heat effects its decomposition into its constituent elements, nitrogen (two volumes) and oxygen (one volume), and to the latter the increased brilliancy is due; mixed with an equal volume of hydrogen or coal gas it forms an explosive compound.

Its most remarkable property (discovered by Faraday in 1823) is, however, that under a pressure of 30 atmospheres, or if exposed to a temperature below − 88° C. (− 126° F.) it liquefies, and may then be preserved in suitable receptacles for any length of time without undergoing change. Thirty gallons of the gas can thus be compressed into about ten ounces by weight of the liquid.

This liquid is colourless, mobile (specific gravity referred to water = ·936), and the least refractive of known fluids; miscible freely with ether and alcohol; water added to it immediately freezes, and in consequence of the latent heat thus set free the fluid gas evaporates, and may be decomposed with explosive violence; a drop of the liquid gas falling upon the hand raises a blister, and the injured surface presents the appearance of having been scalded.

It is very susceptible to slight changes of temperature, one volume of the liquid at 0° C. becoming 1·202 volumes at 20° C. (68° F.); if cooled down below − 115° C. (− 175° F.) it solidifies, and this temperature is often produced when the liquid is allowed to escape from a small orifice, and the solid particles may then block up the exit tube; if the escape is allowed to take place into a suitable metal box, " snow " is formed and may be collected.

Whether as a gas, a liquid, or in solution, it has no action upon vegetable colouring matters (e.g., litmus) exposed to its influence.

PREPARATION.

Nitrous oxide gas is formed when nitric oxide acts upon moist iron filings, &c., or when nitric acid diluted with eight or ten times its volume of water is allowed to act upon pure granulated zinc; for practical purposes, however, it is obtained by heating ammonium nitrate (a crystalline salt deposited on neutralising nitric acid with ammonia or ammonium carbonate); the reaction represented in the following equation then occurs, viz.:—

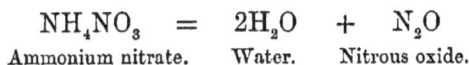

$$NH_4NO_3 \quad = \quad 2H_2O \quad + \quad N_2O$$
Ammonium nitrate. Water. Nitrous oxide.

Figure 1 represents an arrangement for the production of the gas which, as it comes off, is passed through a series of wash bottles; the first (1) is employed to catch the water which comes over with the gas, the second (2) contains a strong solution of sulphate of iron, with which any nitric

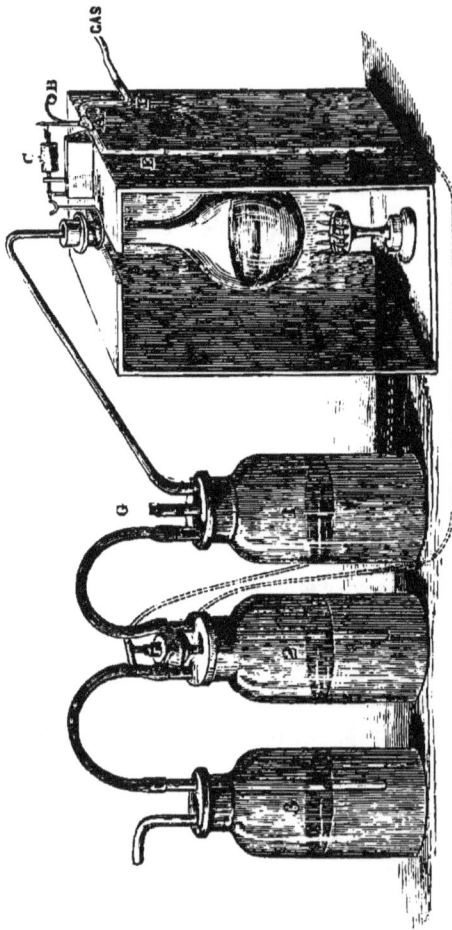

FIG. 1.—Apparatus for the preparation of Nitrous Oxide. The Ammonium Nitrate is placed in the flask and heated by the burner below. A, B, C, arrangement for regulating the supply of coal gas; D, E, pipe supplying coal gas to the burner. 1, 2, 3, wash bottles.

oxide or other of the oxides of nitrogen that may be present combines, and the third wash bottle (3) contains caustic potash for the removal of traces of chlorine, and the neutralising of any free acid; the addition of one or two more bottles containing distilled water is of service in completing the washing.

When the flask and its contents fall below a weight regulated by the counterpoise C, the catch B is released, and presses at A upon the pipe E D, supplying the burner F, and at the same time the flask itself is raised. On the second wash bottle is seen an ingenious mechanism for controlling the supply of coal gas according to the rapidity of production of the nitrous oxide. By these contrivances there is no fear of the gas being evolved at too high a temperature or too quickly, or of the flask being broken.

Before commencing the operation, the purity of the ammonium nitrate should be ascertained, and for this purpose a solution of the nitrate should be added to solutions of chloride of barium (six grains to the drachm), and nitrate of silver (four grains to the drachm), respectively; the formation of even a faint cloudiness or precipitate indicates on the one hand the presence of sulphates or carbonates, and on the other chlorides, and in either case should lead to the rejection of the salt. It is also as well to heat the crude salt gently in a saucer, in order to expel the moisture, and allow it to recrystallise before powdering and introducing it into the flask.

Great care should be taken with regard to the temperature at which the distillation is conducted; this should not exceed at any time 240° C. (464° F.), if higher a complicated reaction takes place and the gas evolved is no longer pure nitrous oxide, while at 315° C. (600° F.) the decomposition is attended with violent explosions. One pound of the crystals yields rather more than twenty-four gallons of the gas.

After allowing for the expulsion of all the air from the apparatus, a point ascertained by the readiness with which the gas issuing from the exit tube re-kindles a glowing chip of wood, the gas is collected in a gasometer similar to the one depicted in Fig. 23, the exit tube of the apparatus being substituted for the bottle. .The water in the well of the gasometer should, for reasons previously explained, be either warm or strongly salted, and it should not be changed oftener than absolutely necessary for purposes of cleanliness, so that it may remain saturated with gas; the addition of a small quantity of fresh water or brine from time to time will be sufficient to counterbalance evaporation.

It is now, however, very seldom prepared at home, though occasions may arise when it may be necessary to do so, and one should therefore be thoroughly conversant with the method of procedure. As a rule, if it is possible to obtain a supply of ammonium nitrate, and the necessary special apparatus, it is equally possible to obtain the compressed gas from one of the many instrument-makers, and at comparatively small cost.

This liquid form is obtained by using a force pump similar to the one used for condensing carbonic acid gas, and known as Natterer's pump, by which the gas from a gasometer or reservoir is driven into stout iron or steel bottles, the heat evolved in the process of condensation being counteracted by surrounding the bottles with a freezing mixture. The amount of gas thus forced in is carefully measured, use being made of the principle of accurately weighing the bottle before and after the operation, or at any time at which it may be desirable to know how much of the liquid it contains; as a rule the bottles are not much more than half filled, in order to allow for the expansibility of the gas when the temperature rises.

ADULTERATIONS AND IMPURITIES.

The simplicity of its manufacture, and the cheapness of the drugs employed in its preparation, render it less liable to adulteration or impurity than might otherwise be the case. It may, however, occasionally present traces of nitric oxide, and other oxides of nitrogen, owing to the distillation of the ammonium nitrate having been allowed to proceed too rapidly or to be conducted at too high a temperature; or it may contain traces of chlorine from the presence of sal ammoniac (ammonium hydrochlorate) in the ammonium nitrate. The presence of these impurities may be suspected if the gas has an irritating or suffocating taste and smell or causes undue coughing on inhalation, and if the gas thus suspected be passed by means of a glass tube through a solution of nitrate of silver, a precipitate will indicate that a chloride is present, and inasmuch as the other gaseous oxides of nitrogen usually yield yellow or orange fumes upon being brought into contact with air, their presence is easily ascertained.

The gas may also occasionally have a rancid, disagreeable odour, due to the decomposition of the oils used to lubricate the condensing machinery and taps.

It is perhaps needless to remark that any of these impurities, even if only suspected, should lead us to reject the use of the particular bottle or sample.

HISTORY.

The discovery of nitrous oxide as a separate chemical compound, is ascribed to Priestley about the year 1776; for the next thirteen or fourteen years it was looked upon simply as a curiosity, and little or nothing was known of its anæsthetic properties. It was re-examined by Sir Humphrey Davy (then Mr. Davy and living at Bristol) about the year

1799, who, although not going so far as to submit to any surgical operation while under its influence, inhaled the gas freely to assuage the pain consequent upon cutting a wisdom tooth ;* but although Sir H. Davy very shrewdly raised the question of its possible use as an anæsthetic in surgical operations, the gas was better known from the excitement it produced, and hence the term laughing gas, which was usually applied to it. In the experiments as then conducted, the gas was simply inhaled from a bag or bottle, through a tube held between the teeth, and no precautions were taken to exclude air.

Thus, although the anæsthetic properties were fairly established, the difficulties in the way of their application, on account of the intensity of the preliminary excitement, precluded the possibility of its practical use.

Matters remained in this state for the next forty-four years. In 1844, Mr. Horace Wells, a dentist of Hartford, Connecticut, was again induced to try its effect in dentistry, and at first with considerable success, but in consequence of one or two failures this success was not permanent, and its use did not become general, more especially as men's minds were at that time much occupied by the discovery and application of chloroform and ether for anæsthetic purposes. But although placed in the background, it was by no means discarded. In 1848 the first surgical operation under its influence (an excision of the breast) was performed by Dr. H. J. Bigelow, and is reported in the "Boston Medical Journal," Vol. I, p. 17.

In 1867, Dr. G. Q. Colton (an American dentist, and for many years previously its most earnest advocate in the States) proceeded to Paris, and mainly in consequence of the

* Collected Works, Vol. III, p. 276.

influence of Dr. Evans, an American practising dentistry in that city, the gas was thoroughly brought to the notice of French medical men and dentists : Prétcrre's advocacy of its claims in 1863 not having met with much success.

In 1864 were made the researches of Hermann, followed by those of Krishaber in 1867; both these authorities speak doubtingly of the gas, and put forward arguments for considering it dangerous if inhaled pure.

In England, Mr. S. Lee Rymer had experimented with nitrous oxide about the year 1864, but no practical results appear to have followed.

In 1868 Dr. Evans came to England, bringing with him Dr. Colton's apparatus, with which he demonstrated at the Dental Hospital (March 31st, 1868), and, in consequence, the Odontological Society of Great Britain appointed a Committee to consider the subject; this Committee was nominated on April 6th, 1868, and presented two very exhaustive and able reports (one the same year, the other in 1872), in favour of its general adoption and recognition as a valuable anæthestic. On these reports, especially the first, and the suggestions made in the discussions which followed them, the principles of our present methods are based. To Mr. Coleman and the late Mr. Clover especially, the profession and public generally are indebted for the earnestness with which they advocated the claims of the gas, both before and after the adoption of the reports above mentioned ; and to their skill in its administration and the ingenuity they and others brought to bear in the construction of suitable instruments, the popularity which nitrous oxide at present enjoys is largely due.

The first practical application of the property of liquefaction is doubtful; it is known at any rate that Dr. Evans brought to this country, bottles containing the gas in a liquid

state, and that Messrs. Barth and Messrs. Coxeter very soon
after prepared a condensed form in bottles very similar to
those at present employed.

No allusion has been made in the preceding pages to
modern workers in the field, from whose writings much of
the information contained in the first portion of this work
is derived, and to whom, therefore, I shall have frequent
occasion to refer.

CHAPTER II.

PHYSIOLOGY AND PATHOLOGY.

OUR knowledge upon these points is derived from two sources, viz. :—

(1) Observations made upon the lower animals in the laboratory both during inhalation and after death.

(2) Clinical observations upon actual patients, combined with the results of post-mortem examinations upon the few recorded cases of death during administration of the gas.

With regard to the clinical observations upon patients, it must be borne in mind that the duration of nitrous oxide anæsthesia is exceedingly brief (see p. 70), and so much has to be thought of and done by both anæsthetist and operator during this short period, that such observations, unless confirmed by further research in the laboratory, are likely to be very misleading.

GENERAL PHYSIOLOGICAL EFFECTS.

Animals suddenly placed in an atmosphere- of pure nitrous oxide very quickly fall down insensible; their limbs twitch convulsively, their breathing becomes panting and stertorous, sensation is completely abolished, and soon the respiratory movements become slow and finally cease altogether. The heart continues to beat for some little time after the

cessation of respiration, and if, before the latter stops altogether, the animal be brought into the pure air, complete
recovery takes place within a very short time, or if the
respiratory movements have ceased, they may generally be
restored by artificial respiration, providing, of course, that
the heart continues to act; further, the same animal can be
subjected again and again to the experiment, at short intervals,
without apparently suffering any injury. If not restored to
air, the heart's action gradually ceases, and death takes place
quite quietly, without convulsions of any sort. As might be
expected, death ensues the more quickly as the activity of
the circulatory and respiratory functions is more pronounced,
and hence in birds it is very rapid, in frogs very slow. It
may be observed in this connection that seeds will not
germinate in an atmosphere of pure nitrous oxide, or if
germination has already commenced it will be arrested, but
is capable of being renewed on the admission of small
quantities of oxygen or atmospheric air; plants, too, cease
to eliminate carbonic acid and do not increase in size.

Post-Mortem Appearances.

In animals dying in a closed receiver after prolonged inhalation of the gas, the lungs are found to be of a light pink
or rose colour, moderately crepitant, and present on their
posterior surfaces one or more small circular well-defined
ecchymotic spots (*ecchymoses sous pleurales*). The blood which
escapes from incisions made in the lungs is more or less full
of gas bubbles, and these bubbles mixed with mucus are also
found in the bronchioles. The right cavities of the heart and
the systemic veins are enormously distended, the left side
and systemic arteries nearly empty. The blood is fluid and
quite black in both veins and arteries. In fact the post-
mortem appearances are precisely those of asphyxia. If

arrangements have been made for the removal of the products
of respiration, signs of simple syncope prevail, with contracted
heart and emptiness of *all* vessels connected with it. In
neither case are any specific or pathognomonic signs of
nitrous oxide poisoning visible.

General knowledge such as is contained in the preceding
pages dates from the time of Sir Humphrey Davy, but we
are now in a position to examine a little more critically the
effects of inhalation upon the various functions and organs
of the body.

SPECIAL PHYSIOLOGY AND PATHOLOGY.

Respiratory System.

1. *Rate and Rhythm.*—With the first few inspirations
the respiratory rate appears to be increased, but this is
undoubtedly due in part, if not entirely, to involuntary
nervous excitement on the part of the patient or animal, but,
excluding as much as possible this source of error, it is found
that the movements of respiration are first increased in
number but not lessened in depth ; they then become slower
than normal, slightly deeper, and accompanied with stertor ;
gradually they appear to stop, but simple pressure on the
chest-wall will often cause them to appear again and
continue for a brief period; finally, even this power of reproduction is lost, and the respirations cease altogether. It
is to be noted that this alteration in the respiratory rate is
not due to any excess of expiratory over inspiratory effort, or
vice versâ, but to a change in the movement in its entirety ;
nor is it accompanied by dyspnoeic convulsions.

2. *Gaseous Interchange.*—In ordinary aërial respirations
we find that the expired air is warmer, moister, contains
more carbonic acid and waste products, and less oxygen than
inspired air. What, if any, corresponding changes occur

when nitrous oxide is substituted for air? Unfortunately, our information upon these points is far from complete; partly on account of difficulties connected with analysis, but mainly because of the impossibility of removing at once all the air that may remain in the trachea, bronchial tubes, and lungs before inhalation commences.

The temperature, degree of moisture, and amount of the waste products in expired gas do not appear to have been determined, such analyses as have been made being directed towards the estimation of the gaseous elements alone.

Professor Frankland, at Mr. Coleman's suggestion, made some analyses in 1869, and the results were published in St. Batholomew's Hospital reports for that year* and elsewhere.

TABLE I.

Gas.	Before Inspiration.	After first Expiration.	After 3rd Expiration.
Carbonic Acid	·103	3·187	2·346
Oxygen	1·540	2·700	1·621
Nitrogen	6·160	17·854	17·100
Nitrous Oxide	92·197	76·259	78·933

The above table represents the result of these analyses; in the first column of figures is the composition of the gas inspired, the second and third columns show the analysis of the gas after the first and third expiration respectively. It should be noted with regard to these figures that they are merely averages, and not obtained from single experiments nor from any one individual; they possess this advantage however, that they are obtained from human beings and not animals.

However imperfect these figures may be, they are

* Art. XIV, p. 153.

sufficiently precise to prove that, partly as a result of simple admixture of gas and air in the bronchial tubes, partly owing to diffusion from the residual air in the air vesicles, and partly on account of the gaseous interchange in the blood to which we shall subsequently refer, the expired gas tends to become precisely similar in composition to that inspired, showing at any rate that no very active decomposition of the gas occurs.   Moreover, the substitution of nitrous oxide for the other gases does not take place at all equally, but mainly at the expense of the oxygen which is rapidly reduced.   The carbonic acid, instead of being increased in quantity, as is the case in ordinary respirations, is rapidly and continuously diminished.   The effect upon the nitrogen is very slow, owing probably to the fact that it is removed simply by diffusion, and does not even under ordinary circumstances take any very active part in the physiological functions.

Dr. Amory's experiments upon himself and upon dogs* gave very much the same result, but were carried somewhat further.   He found that the relative amount of carbonic acid exhaled was only about half of that found in the same number of aërial respirations.   The details of the methods he employed and the figures upon which this conclusion is based are too complicated for reproduction here, but inasmuch as the experiments were made upon the same animal or individual, the results may be looked upon as even more reliable than those of Mr. Coleman above quoted.

As might be expected, the elimination of carbonic acid for some hours after recovery is in excess of that observed in health.

* "New York Medical Journal," August, 1870.

Circulatory System.

1. *The Heart.*—Cardiographic observations have not proved at all satisfactory ; judging, however, from the pulse it would appear that the action of the heart is at first distinctly accelerated, while at the same time it loses somewhat in force ; but precisely similar reservation must be made in this as in the case of the apparent increase in the respiratory rate, *i.e.*, that this increase is partly or wholly emotional. As the

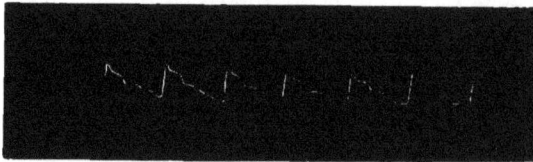

FIG. 2.—Sphygmographic trace of pulse before inhalation.   (Dudley Buxton.)

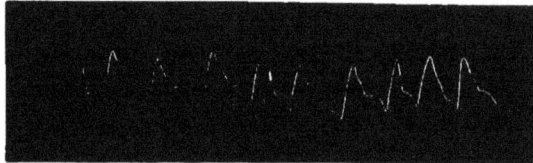

FIG. 3.—Tracing of same pulse while under gas, showing acceleration, loss of
tidal wave, and accentuation of dicrotic wave.

FIG. 4.—Same pulse, patient gradually coming round, showing increased
firmness of heart-beat.

patient or animal passes more fully under the influence of the gas, the heart becomes quieter, and if anything firmer in tone, but it still beats quicker than usual; if the gas is pushed the beats first become slower, then intermit and may finally cease altogether. It continues to act, however, for some little time after the respirations have altogether ceased, and as long as it does so the animal is usually capable of resuscitation by means of artificial respiration.

The pulse tracings shown in Figs. 2, 3, and 4, for the use of which I am indebted to the courtesy of Dr. Dudley Buxton, are extremely interesting, and indicate precisely the changes above alluded to, *i.e.*, primary acceleration with increased firmness as shown by the greater sharpness of the initial curve.

2. *The Vascular System.*—The vessels at first appear to be little if at all affected; at a later stage, however, the peripheral vessels dilate (Buxton), with the exception of those of the splanchnic area (kidney, spleen, &c.), which contract in a very decided manner, contrasting in this respect with the vascular changes observed in asphyxia, when due to deprivation of air combined with accumulation of waste products in the lungs. It follows from this dilatation that the blood current in the peripheral vessels themselves and in the capillaries is slowed, and a certain amount of blood-stasis is produced; but this slight congestion is quite secondary and in no way sufficient to account for the very decided lividity almost invariably observed during inhalation.

3. *Blood Pressure.*—A further study of the sphygmographic tracings on the preceding page shows us that the tidal or first of the descending waves in the normal pulse, gradually disappears or is reduced to the merest indication, the dicrotic or second wave becomes at the same time very marked, and further removed from the apex of the trace; these points associated

C

with the knowledge that the systemic vessels are dilated would suggest a lowering of the arterial pressure. Dr. Dudley Buxton,* to whose admirable papers upon the " Physiological Action of Nitrous Oxide " I am indebted for many of the facts contained in this chapter, has demonstrated by experiment upon animals that during the first stage of inhalation little or no change in blood pressure occurs, but that very soon a slight fall is observed ; he has further proved that after recovery the blood pressure rises slightly above the normal, not at once, but by a series of irregular curves, and that this elevation persists for some little time.

4. *The Blood.*—The most obvious effect upon the blood as a whole is, that it becomes darker in colour, in other words retains its venous characteristics ; to this, rather than to the stasis, is due the lividity of the skin and mucous membranes, the darkened blood circulating in both arterial and venous systems; it is highly probable, too, that the nitrous oxide exists in the form of loose chemical combination with one or other of the constituents of the blood, but with which is quite uncertain. Some observers have described certain spectroscopic changes, but this has been denied by others and requires confirmation.

Under the microscope the corpuscles of human blood show no change ; those of the frog, observed while still in the vessels, are said to become slightly flattened.

Chemical analyses of the blood of dogs made by Drs. Jolyet and Blanchet† yield the following interesting results, the figures representing the percentage of the various gases :—

---

* "Transactions of the Odontological Society," 1887.
† "Archives de Physiologie," July, 1873.

TABLE II.

	Breathing Air.	Breathing Nitrous Oxide.		
		105 Seconds.	3 Minutes.	4 Minutes.
Carbonic Acid ..	48·8	37	36·6	34
Oxygen	21	5·2	3·3	·05
Nitrogen.. ..	2	·7	nil	nil
Nitrous Oxide ..	nil	28·1	34·6	37

From this table we see that—

1. The amount of carbonic acid in the blood slowly diminishes as inhalation proceeds.
2. The oxygen is rapidly reduced to a mere trace.
3. The nitrous oxide gradually increases in quantity, taking the place of the other gases, but especially of the oxygen.

In fact, precisely similar changes as have been shown to occur in the character of the expired gas. A strict comparison of the above figures with those on page 14 is hardly possible, as the analyses and experiments were not made by the same observers, nor under the same circumstances ; but by comparing results we are, I think, entitled to conclude—

1. That the progressive loss of carbonic acid observed in expired gas is not associated with any accumulation of that gas in the blood, but on the contrary with diminution, and hence such loss is probably due, either to lessened production in the tissues, or to defective absorption by the blood.
2. But in order that carbonic acid may be produced in the tissues, we know that free oxygen, or an agent capable of yielding up its oxygen must be present

c 2

in the blood. But the free oxygen is displaced by nitrous oxide, and the latter undergoes no decomposition or any alteration whatever in the blood.

3. Hence it appears probable, that during inhalation of the gas the process of tissue metabolism is in abeyance.

Nervous System.

1. *Physical Changes.*—The effects of nitrous oxide upon the physical condition of the cerebral system were first observed by Dr. Amory (*op. cit.*), and his observations have been confirmed and extended by Dr. Dudley Buxton (*op. cit.*).

(*a*) The cerebral pulsations at first increase and then diminish in number, apparently in direct proportion to the variations in the respiratory rate.

(*b*) The whole cerebro-spinal system increases in size.

(*c*) This increase in size is accompanied by, and is probably due to, a dilatation of the contained blood-vessels and slowing of the blood current.

2. *Functional Changes.* (*a*) *Brain.*—There is first a period of excitement and hyperæsthesia, associated with exaggeration of auditory and visual sensations, and some mental exaltation, hence pleasing dreams, rhythmic movements, &c. ; then perversion of intellectual and moral sense, hence erotism and hysterical phenomena. This is quickly followed by a condition of narcosis gradually increasing in depth, hence inability to originate muscular movements, loss of appreciation of tactile and painful sensations, abolition of inhibitory power, and if the gas is pushed, paresis of the centres in the medulla presiding over vital functions, with consequent cessation of respiratory and cardiac action.

(*b*) *Spinal Cord.*— As in the case of the cerebral functions

so with the spinal, a double action is probably produced, first
of excitement or hyperactivity, then sedative; but if we
consider the threefold part which the cord plays (as a centre
for reflex action, as a special centre, and as a conductor), to
say nothing of the nervous idiosyncrasies of the patient, we
can readily understand how difficult it is to differentiate
these actions, and hence the apparent confusion and irre-
gularity in the development of the symptoms observed.

(α) *As a centre for reflex action.*—The skin or superficial
reflexes (*e.g.*, conjunctival) are abolished early, their dis-
appearance being probably preceded by a period of slight
exaggeration: of the deeper reflexes, the patellar usually
persists, and ankle-clonus (absent in health) is sometimes
developed.* The tremors, twitchings, and convulsions are
probably due to an exaltation of these deeper reflexes, to
which also may be ascribed the spasm, oposthotonous, pleurosthotonos, emprosthotonos, and irregular muscular movements
occasionally observed.

(β) *As a special centre.*—The involuntary defæcation and
micturition, which sometimes takes place during the latter
stages of inhalation, seem to be due to paresis of the special
spinal centres which preside over these functions.

(γ) *As a conductor.*—Many of the above symptoms may
also be referred to an alteration in the conducting power of
the cord, so that inhibitory stimuli from the brain are no
longer transmitted, but the exact nature and cause of this
alteration is very uncertain.

Quite conjectural also are the cause and explanation of
the rare secondary symptoms, *e.g.*, prolonged coma, hemiplegia,
and paraplegia.

(c) *Special sense.*—Of the organs of special sense the only
phenomena to which allusion has not been made, and which

* Dr. Buxton, "British Medical Journal," September 25th, 1887.

seems to be of importance, is the dilatation of the pupil. This, under ordinary circumstances, may be due to paresis of the motor-occuli nerve, with consequent loss of power in the circular muscular fibres of the iris; to irritation of the sympathetic, producing increased action of the radiating fibres; or to the action of a special centre presiding over dilatation, and which is said to exist in the medulla; it may also be noted that violent muscular efforts are known to be associated with dilatation, hence it may, in the case of nitrous oxide inhalation, be due to the muscular spasm. Much more likely than all these, however, is its connection with vaso-motor disturbance : and it should always be borne in mind, that sudden dilatation invariably precedes or is associated with syncope.

But few words are necessary as to the effects of the gas upon other functions, as but little is at present known upon the subject.

The effects, if any, upon the secretions of the liver and kidney are unknown. M. Laffont[*] has referred to the increase of sugar in the urine of glycosuric patients, but the point requires further elucidation.

In conjunction with the digestive system, passing allusion must be made to the development of retching and vomiting which is probably purely reflex, due in part to mechanical irritation of the back of the tongue and fauces, and hence should more strictly be referred to the nervous system : borborygmi which are so frequently heard may be accounted for by the alteration induced in the splanchnic vascular area, or, as is much more likely, to simple nervousness on the part of the patient prior to taking the gas, actual increase in the intestinal peristalsis apart from this has not been noted. These phenomena (retching, &c.) have also been ascribed to

* "Comptes Rendus, Société de Biologie, Paris," Vol. XII, No. 37.

mechanical distension of the stomach, consequent upon swallowing the gas, but of course the process of deglutition is hardly possible if the mouth is held widely open with a mouth-prop; at any rate, the amount of gas swallowed under such circumstances must be quite insignificant: the mere attempt at swallowing with the mouth in such a position is in itself likely to induce retching.

CHAPTER III.

Mode of Action.

Under the head of Physiology we have pointed out the various physical and psychical effects observed during administration, and have at the same time suggested the proximate or more immediate causes of the several special symptoms. No attempt was however made, to show the relation between these various symptoms, as will be done in discussing the clinical phenomena, nor to explain the *modus operandi* of the gas in the production of its, to us at any rate, most important phenomenon, *i.e.*, the anæsthesia: this latter especially is a point of considerable interest to all who study the subject, and of no little importance to the practical anæsthetist, and is deserving of our best attention. Taking into consideration its somewhat theoretical nature, it is not surprising that the matter should have led to much controversy and dispute, and that the difficulties met with in attempting to explain clearly the views of the several exponents have been considerable.

At one time it was even suggested, that the peculiar anæsthetic effects of the gas were due to its action upon the nerve trunks and ends. That its action is upon the great nerve centres themselves is now generally admitted, and was formerly explained on the supposition that the gas was decomposed, either in the blood or in the tissues, into its constituent elements, nitrogen and oxygen; as the relative amount of oxygen (36 per cent.) would then be greater than that contained in atmospheric air (21 per cent.), the theory was put forward that the phenomena observed are those due to hyper-oxygenation, others ascribing them to the presence of a large amount of free nitrogen. Inasmuch, however, as this decomposition was shown not to occur, the theories in question are no longer held. Both Hermann (1864) and Krishaber (1867), two of the earliest investigators, looked upon the gas as practically irrespirable if pure, considering that whatever clinical success had hitherto attended its use, was due rather to its impurity and admixture with air or oxygen, and that it was certainly a dangerous agent. But in spite of these and other adverse opinions the popularity of the gas increased and is still increasing, leaving the success, safety, and exact mode of action to be explained as they may by theorists and physiologists.

The question now rests between those who hold the theory of asphyxia and those who maintain that the gas possesses specific anæsthetic properties.

According to the former, the gas is actually irrespirable in itself when chemically pure, or at best acts merely as an indifferent or innoxious agent, taking the place of air, and producing the symptoms observed during its use by a process allied to asphyxia, i.e., by depriving the tissues of oxygen, and allowing the carbonic acid and other waste products to accumulate. In support of this theory, it was urged that the symptoms observed during inhalation were very similar to those preceding death by asphyxia, and no doubt this was the case in early days, when the gas was simply breathed in and out of a bag, and no arrangement was made for getting rid of the products of expiration. When, however, by the provision of suitable valves, the gas coming from the lungs was no longer permitted to mix with that from which the supply was drawn, and it was also proved experimentally that the carbonic acid did *not* accumulate in the blood, a modification of this theory was necessary, and it was then suggested that a condition of asphyxia still existed, but only partial asphyxia due to deprivation of oxygen, but that even such partial asphyxia was sufficient to account for all the symptoms.

Dr. Amory in America (*op. cit.*), and Dr. George Johnson in England,* as the result of very careful experiments and observations, put forward able arguments in support of this view, and in explanation of the changes which occur during inhalation. The summary of their conclusions, as revised by Dr. Johnson himself, is as follows, viz. :—

1. The oxygen in the lungs and blood is replaced by nitrous oxide, which does not itself undergo any chemical change.

2. The black unoxygenated blood excites contraction in the muscular walls of the systemic arterioles, with consequent fulness and high tension of the pulse.

3. That the same condition subsequently occurring in the pulmonary arterial system leads to corresponding stasis in the pulmonary capillaries, and hence emptiness of the systemic arteries, venous congestion, and feebleness or loss of pulse in the latter stages.

4. That the anæsthesia is due to insufficient oxidation of the nervous structures, and the convulsions and twitchings are precisely similar in nature and origin to the epileptiform seizures following extreme cerebral anæmia, *e.g.*, after ligature of a vessel.

Further, Messrs. Jolyet and Blanche (*op. cit.*), as the results of their experiments, concluded that chemically pure

---

* "Medical Times and Gazette," April 3rd, 1869. See also Dr. Johnson's "Medical Lectures and Essays," Chapter II.

gas supported neither vegetable nor animal life and was therefore irrespirable : that anæsthesia only ensued when the amount of oxygen in the blood fell to 2 or 3 per cent. (the normal amount being 21 per cent.), and that it was to this latter fact that the anæsthesia was due, and they go so far as to maintain that this being the case, the use of the gas should be proscribed.

Those that maintain that the peculiar effects are due solely and entirely to its specific properties point out—

1. That the similarity which exists between the clinical symptoms of nitrous oxide inhalation and asphyxia, is, after all, rather apparent than real. There is none of the violent effort to obtain breath, no reflex sighing and yawning, no feeling of constriction, no vertigo or dimness of vision in the former as in the latter.

2. In asphyxia, physiologically considered, the course of the symptoms is as follows :—first, true dyspnœa, then violent efforts at expiration with expiratory convulsions, and finally exhaustion ; at the same time the blood pressure at first steadily rises, the heart beats more quickly and more forcibly, and it is not until exhaustion sets in that any fall in pressure or pulse rate is noted. In nitrous oxide inhalation, on the other hand, almost diametrically opposite phenomena are observed. The dyspnœa is simply rapid (not forcible) breathing, no excess of either inspiratory or expiratory effort is observable, and no truly dyspnœic or expiratory convulsions ensue ; the systemic arteries dilate, the blood pressure falls, and the heart-beat, when once the previous mental excitement has been overcome, slowly and steadily decreases in force and frequency.

3. The condition of anæsthesia is induced at a very much earlier period, both relative to the development of other symptoms and absolutely, than in any but the severest forms of asphyxia.

4. In asphyxia the cerebro-spinal systems decrease in size, in nitrous oxide inhalation they distinctly increase in volume, and with regard to the splanchnic vascular areas, the opposite differences are observed.

While admitting that a very marked dissimilarity may exist between the phenomena observed during the two conditions (asphyxia and inhalation), it would be perfectly compatible with all we know of either state to attempt to explain the symptoms following nitrous oxide administration on the supposition, that they are due partly, and in the first instance, to the specific action of the gas, and partly, in later stages, to absence of oxygen.

The late M. Paul Bert proved this theoretical position by a series of crucial experiments, which subsequently stood the test of actual practice. He found that anæsthesia only ensued when the amount of gas in the blood reached 45 per cent. ; that any attempt to obtain a higher percentage was followed by death, or the development of serious symptoms, in consequence of the failure of the oxidising processes in the tissues ; that if it was attempted at the ordinary atmospheric pressure to supply the necessary amount of oxygen, the percentage of nitrous oxide sank below that requisite for maintaining anæsthesia, and the patient or animal consequently recovered consciousness. Arguing from these premises, he suggested that if means could be found to supply enough oxygen to maintain vitality, without diminishing the relative amount of nitrous oxide in the blood, all the advantages of nitrous oxide anæsthesia, without the subsequent partial asphyxia, would follow. But at the ordinary barometric pressure, any admixture of the two gases leads to a relative diminution in the amount of nitrous oxide available, and hence to imperfect anæsthesia.

Obviously, the only way was to arrange for the inhalation of a mixture of nitrous oxide and oxygen under pressure, so that while the same absolute amount of nitrous oxide is absorbed, its diminution in volume, consequent upon increased pressure, is made up for by the addition of definite proportions of oxygen. Experiments upon animals in this direction were perfectly successful, and by the use of an iron chamber devised by M. Fontaine, of such size as to be capable of containing patient, operator, administrator, assistants, nurses, &c., and so constructed that the pressure within could be varied to a nicety, it was at length (February 13th, 1879), found possible to perform an operation of some duration upon a patient while perfectly anæsthetised by means of a mixture composed of nitrous oxide (85 per cent.), and oxygen (15 per cent.), the whole being under a barometric pressure of 92 ccm. (normal = 76 ccm.). Since that date many major operations have been performed in the chamber, the patient being under the influence of this mixture, and although of late years the practice has rather fallen into disuse, this is owing rather to the lamented death of its originator, and to the cumbersomeness of the apparatus, &c., than to any failure in the actual process, or fault in the argument. Without being so sanguine as to expect that the time will ever arrive when it may be possible, for the sake of the anæsthetic alone, to conduct all operations in specially constructed chambers, either in hospitals, or movable from house to house on wheels as his followers would suggest, but at the same time bearing in mind the ridicule showered upon, and the arguments once urged against the now almost universal system of antiseptic surgery, it may not be altogether absurd to look forward to a practical development of the plan above mentioned. Sub-sequently the same investigator proposed to administer the gas very freely at first, and to prolong the anæsthetic action by means of a mixture of oxygen and nitrous oxide in definite proportions under the normal pressure. Experi-mentally this plan was found to be of some value, but it does not seem to have been carried beyond this stage.

The course of our inquiry into the subject of the *modus operandi* of the gas has therefore led us to the following conclusions, viz. :—

1. The gas undoubtedly possesses in itself specific anæs-thetic properties.

2. These specific properties require for their satisfactory development the presence of very large quantities of gas in the blood.

3. Hence the gas must be given pure, to the exclusion of all traces of atmospheric air.

4. That this exclusion of air leads to the development of symptoms of oxygen starvation (not true asphyxia).

5. That as at present administered it is impossible to exclude these symptoms, though it has been done by the special methods of M. Paul Bert.

CHAPTER IV.

APPARATUS.

FOR the ready and successful administration of nitrous oxide, certain special forms of apparatus are necessary: these may very readily be described under four heads, viz. :—

I. The apparatus connected with the storage of the gas.
II. The conducting apparatus, or that used in conveying the gas to the face-piece.
III. The face-piece.
IV. Accessory instruments, gags, forceps, &c.

STORAGE.

As previously mentioned, the gas is usually obtained from the instrument-makers in the liquid form, and is then contained in strong metal bottles made of wrought iron, or of steel, the latter being equally strong, much smaller, and lighter; any size bottle can be obtained, but those most frequently used are calculated to hold $7\frac{1}{2}$, 15, and 30 oz. by weight of the liquid respectively, corresponding to 25, 50, and 100 gallons of the gas itself.

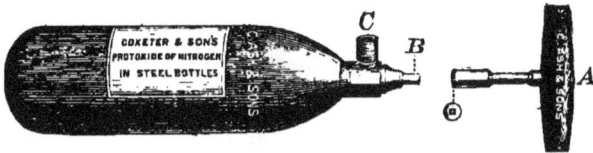

FIG. 5.—Steel gas bottle and key.

Fig. 5 represents a 50-gallon bottle; at B is a powerful valve on turning which the gas escapes from the orifice C, which is furnished with a screw for attachment to the gasometer, or to the metal union of the conducting tube. At A is a handle or key for turning the valve, another shape being represented on page 56 (Fig. 23–4). A pedal key for use with the foot, the bottle being fixed in a suitable support, is shown in Figs. 6 and 7. Each bottle has affixed to it a label, recording its weight when empty, and when sent out from the instrument-maker's, from which it is easy to deduce the amount of gas at any particular time, remembering that each gallon of gas is condensed into $\frac{3}{10}$ of an ounce of liquid.

As a matter of convenience it will be found as well to work with a pair of bottles, so that one is constantly full while the other is in use, and there is then no danger of sudden failure of supply at a critical moment.

It may not be out of place here to suggest a few precautions in dealing with the gas in this form.

(1) Keep the bottles as far as possible at an equable temperature, recollecting its susceptibility to changes (page 3); do not at any rate place them in hot water, or near or in the fire, as it is recorded has been done by more than one ingenious individual, with astonishing and sometimes fatal results.

(2) Be careful in handling the bottles while in use or if they have recently been used; the cold produced by the conversion of the liquid into gas is intense, the cold metal readily blistering the fingers.

(3) Weigh the bottles yourself as received, and from time to time, instrument-makers may be careless, and taps and valves may leak.

(4) Over-filling of the bottles may be suspected when, on turning the valve, the gas escapes in an irregular spasmodic manner, and the result of this defect is, that frozen particles of liquid choke up the orifice C, or may be forced into the conducting tube or

reservoir bag (if the direct method of administering is used) and may by their subsequent sudden expansion, cause explosions, &c.

The gas contained in the bottles above described, may be either administered to the patient through the medium of a suitable conducting apparatus, (the direct method), or it may be passed first into a gasometer similar to the one shown on page 56, the mode of action of which is precisely similar to that of the ordinary gasometer used, on a larger scale, for the storage of coal gas—i.e., a metal reservoir (1), sinking into a well or tank of water (2), and counterpoised by weights (5) passing over the pulleys (6). Gas is admitted into the reservoir by connecting the tube with the gas bottle (3), or with the apparatus used in the manufacture of the gas. At B is the tube for conveying the gas from the gasometer to the face-piece (D).

This gasometer may be kept in a room immediately beneath or adjacent, and the tube B passed through the ceiling or wall to a stand pipe on the floor of the operating room, by the side of the chair; or it may be placed permanently upon a shelf in the operating room, as close as possible to the patient; or it may be fitted with castors, and only brought forward from its cupboard or recess as occasion requires. If kept in another room it is usual to have a dial plate or tell-tale of some sort fitted in the operating room and connected with the gasometer, so that the amount of gas in the reservoir may be ascertained at any moment, and without leaving the room. A cord passing over a pulley, and carrying a weight which works up and down against a fixed index on the wall, is the simplest form of tell-tale.

The main advantages in working with a gasometer are, I believe—

1. That the gas, always being under slight pressure, is

D

forced continuously and evenly through the tubes
and face-piece.

2. That in event of inability or neglect to turn off the
valve of the gas bottle completely, there is less
waste, the gas simply flows into the reservoir and
can be used at any time.

Other advantages have been claimed for it, but they are
possessed in equal measure by the more direct method, and
against those enumerated we have to place the following disadvantages :—

1. The cost and cumbersomeness of the apparatus, and
the constant attention it requires.

2. The waste of gas in using the apparatus consequent
upon its being forced continuously through the
face-piece, &c.

3. The tendency of the gas, if left standing for any
length of time in the reservoir, to take up odorous
gases from the stagnant water.

The more direct method of administering is rapidly
becoming more general, on account, mainly, of the handiness
and portability of the apparatus, and the slight attention
required to keep it in working order, but also because the
anæsthesia produced from the use of *fresh* gas is more certain
and satisfactory. In this method the supply of gas, through
the conducting tubes and reservoir bag into the face-piece, may
be maintained by means of a key turned by the hand of the
operator or assistant (Fig. 25), the bottle being placed on a
convenient chair, or the valve may be turned by the foot of
the administrator: in this latter case one or two bottles are
fixed either vertically (Fig. 6) or horizontally (Figs. 7 and 24),
and the valve is turned through the medium of a flat pedal
or foot-key, or the valve is fixed, and the bottle itself rolled
backwards and forwards on the floor with the foot.

If the bottles are arranged horizontally the taps should be slightly raised, so that the contained liquid may have a tendency to fall away from the valves and exit tubes. I am

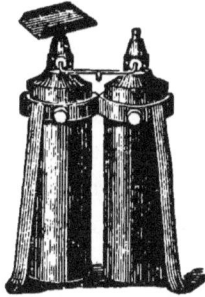

Fig. 6.—Gas bottles in a vertical stand, turned by the foot.

myself in the habit of working with the arrangement depicted in Fig. 7, where the relative position of the valve and exit tube are reversed, and the bottles are held together by a stout tube. which also serves as a handle for carrying, rotation being

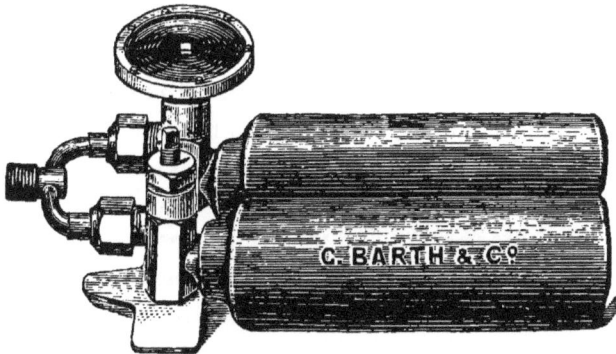

Fig. 7.—Gas bottles fixed horizontally.

prevented by a flat metal plate which is attached by screw nuts to the bottles, and whose under-surface is furnished with two or three short spikes which hold the bottles firmly on the floor, and also serve to raise the ends. The pedal can easily be transferred to the other bottle, should the one in

use run out. With the foot on the pedal, rotation of the heel
to the right opens the valve, and with a very little practice
the amount of gas allowed to escape can be regulated to
a nicety.

Although I would myself advise the use of the direct
method, it must be quite understood that whichever method
is adopted, a thorough acquaintance with the working and
peculiarities of the one chosen fully counterbalances any
supposed advantages possessed by one method over the other,
and that equally good work has been done with both.

THE CONDUCTING APPARATUS.

If the gasometer is used it will be necessary to obtain—

1. An air-tight tube of about 1 inch or more in diameter
   (Fig. 23 B), the length of which must of course
   depend upon the distance of the gasometer from
   the patient, but it should be as short as possible
   compatible with freedom of movement, and in
   addition should be perfectly flexible. This tube is
   attached at one end to the exit tube of the gasometer
   (7), and at the other is fitted to one arm of
2. A metal tube bent at right angles, in the angle of
   which is a stopcock capable of two distinct movements (Fig. 23 C); in the first position the tube of
   the gasometer is shut off, and communication with
   the air alone is permitted, in the second the air-
   hole is closed, and communication with the tube
   maintained. The other arm of this metal tube is
   fixed to a corresponding tube on the face-piece.

A bag, similar to the reservoir bag to be described below,
but smaller, is sometimes inserted in the course of this tube,
but such an arrangement does not seem to possess any very
obvious advantage.

If it is intended to obtain the gas directly from the gas bottle the following apparatus will be required :—

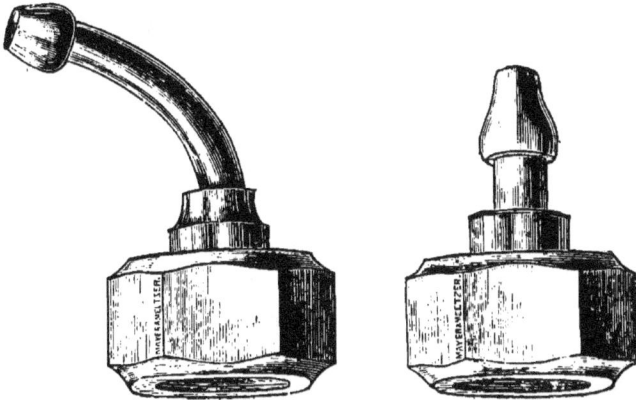

FIGS. 8 & 9.—Metal Unions for attachment to the gas bottle and conducting tube.

(1) A metal union for screwing on to the exit tube of the gas bottle (Figs. 8 and 9).

(2) An extra thick indiarubber tube, the conducting tube, of small diameter (Fig. 24 B), 1½–2 yards in length; this is attached at one end to the nozzle of the metal union, and at the other to

(3) An impervious rubber bag, "the reservoir bag" (Fig. 24 C), of about 2 gallons cubic capacity. The so-called Cattlin's bags are, in my opinion, needlessly thick and heavy, I prefer much lighter ones, nearly as light as those known as "Ether bags." This bag is fitted, either directly or by means of a metal collar, to a two-way metal tube and stopcock similar to that described above, or to one arm of the three-way tube described below: in attaching the above-mentioned conducting tube to this bag, care should be taken that the free end is thrust well into the bag itself, and that the lumen of the tube is not compressed in the slightest by any cord or string

that may be employed to tie the two together; as a matter
of fact it will be found that a simple indiarubber ring or
band will suffice for the latter purpose. If these simple
precautions be taken, the hissing sound produced by the gas
rushing into the reservoir bag is reduced to a minimum, and,
with a little judicious management of the supply from the
bottle, may be completely abolished.

I have here advised, and myself habitually use, this reservoir bag attached to the arm of the angular three-way tube,
as near the face-piece as possible, but it is only right to
mention that some of our best authorities upon the subject of
anæsthetics, employ a tube of large diameter between the
reservoir bag and the face-piece (Fig. 25), in some instances
of sufficient length to hang the bag over the shoulders of the
administrator. This undoubtedly avoids the possibility of the
patient clutching at, or destroying the bag should there be
any struggling or resistance to the administration, but I have
seldom seen the necessity for this, and it undoubtedly adds to
the cumbersomeness and weight of the apparatus, and, the
gas being heavier than air, the longer the intervening tube,
the more difficulty does the patient experience in obtaining a
free and full supply.

If this long tube is employed, or the gasometer used,
another bag attached to the face-piece, and termed the
supplemental bag, is usual, as will be described below.

All bags employed in the inhalation should be frequently
cleaned, especially if much used; this may readily be done
either by detaching the bags from their connections and
turning them inside out, or by inflating them with air, and
rapidly passing into their interior a handkerchief or dry
rag. The tubes and stopcocks should also be cleansed by
passing through them a stream of water, taking care to dry
thoroughly and to lubricate the stopcock before re-use.

THE FACE-PIECE.

This is really the most important part of the whole apparatus, for on the accurate fitting of the face-piece to the mouth and nose, much of the success of the administration depends, too much care cannot be exercised in its selection, and in accustoming one's-self to the use of the particular shape chosen.

FIGS. 10 & 11.—Face-pieces. A, Expiratory valve; B, Inspirating valve; E, Tube for attachment of supplemental bag.

Numerous forms are in use, some made of thin metal covered with leather, others of stiff india-rubber or thick leather; some oval in shape, others conical, &c. (Figs. 10, 11, 12 and 23).

One important point they have in common, which is, that round the oral margin they are all furnished with a hollow pad of rubber which is kept inflated with air through a little stopcock, and which, by its softness and flexibility, materially assists in the close and accurate adaptation to the irregularities of the surface to which the face-piece is applied, a point of no little importance, when the necessity for administering the gas as little mixed with air as possible is borne in mind.

In America it is deemed advisable by some, to cover up as little of the face as possible, the gas being inhaled through a hard rubber mouth-piece held between the teeth of the patient; to this is sometimes added a transverse metal hood which fits over the lips. It is necessary, when this method is adopted, to compress the nose, with the fingers, or by the use of a nose clip, or to adopt some other means to prevent the passage of air through the nostrils. In children, and in some morbid conditions of the jaws of adults, it is not possible to insert such a mouth-piece between the teeth, and face-pieces, similar in principle to the English forms, are rapidly coming into more general use in the States.

I shall only here attempt to describe two varieties of face-piece, viz., those with whose use I am myself most familiar, though not claiming for them any very special advantage over other shapes and makes.

1. Clover's face-piece (Figs. 10 and 23 D) is oval in shape, and made of soft metal or stiff leather, so as to be readily moulded to the mouth and nose; round the edge is the usual air-pad of rubber; on its upper surface, which is flat and of stouter make than the body of the instrument, are two short tubes, one of which (A, Fig. 10) contains a small valve, opening during expiration, and closing by means of a light spring on inspiration; this valve should be readily accessible to the finger of the hand holding the face-piece so that its action may be controlled if necessary. The other tube, B, is the one to which one arm of the two-way tube of the reservoir bag, or the tube of the gasometer, is fitted and is furnished with a valve which only opens during inspiration. There is also frequently a third tube, E, in the anterior portion of the body, to which is attached a straight metal tube carrying a small rubber bag of about $\frac{1}{2}$-1 gallon cubic capacity, and to which the term "supplemental bag" is applied; this bag

should be of much the same material as the reservoir bag, and the tube to which it is attached is furnished with a closely fitting stopcock, by means of which the communication between the bag and face-piece can be opened or closed at will: when the bag is not used, the tube in the face-piece is covered with a tightly fitting metal cap. No valves are attached to any part of the supplemental bag and tube, and the inspiratory valve above mentioned is often dispensed with.

2. The face-piece depicted in Fig. 12 consists of a short cone of stiff rubber with the usual air-pad round the edge; the apex is of metal, into which fits very tightly one arm of a three-way tube, one of the other arms carries the reservoir

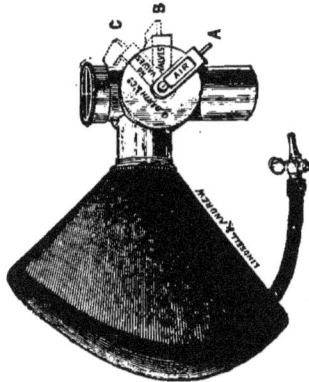

FIG. 12.—Barth's improved face-piece and three-way tube.

bag, and the third arm contains a simple expiratory flap-valve. At the junction of the three arms is situated a stopcock, which also contains a rubber flap-valve, and the movements of the tap are so arranged that on turning it more or less round we obtain—

(*A*) Communication between the face piece and the external air alone, the bag being shut off.

(*B*) Communication between the face-piece and reservoir bag, both valves working.

(*C*) Communications between the face-piece and bag alone, both valves cut off, so that the reservoir bag is converted into a supplemental bag. This arrangement recommended itself to my notice by its extreme simplicity and lightness, and having used it for some time past I can speak confidently of its being quite effectual.

Great attention should always be paid to the cleanliness of the face-pieces, in which saliva, blood, &c., are apt to collect, and they should be washed or wiped out with a wet sponge after each administration.

ACCESSORY APPARATUS.

Under this head I include such smaller instruments as are of use in ordinary administrations, together with those that are of service in cases of difficulty or danger.

1. So-called rarefiers or regulators (Fig. 13), consisting of a

FIG. 13.—Rarefier.

box into which hot water is poured to surround a narrow metal pipe through which the gas runs; they are attached directly to the gas bottle, and are very serviceable in maintaining the equable flow of the gas, and inasmuch as they ensure its full expansion, promote economy.

2. In order to moderate the hissing sound made by the gas rushing into the reservoir bag, so-called silencers or quieters have been introduced (Fig. 14).

FIG. 14.—Silencer.

They consist essentially of a metal tube, about one inch in diameter, and six to eight inches long, and are filled with small particles of cork, sponge, glass, &c. They are attached by one end to the gas bottle, and by the other to a wide flexible tube leading into the reservoir bag. Objections have been raised to them on account of the amount of air contained in the wide tube, and which it is necessary to expel before commencing the inhalation. Personally I have not much experience of their use, as I have never had any objection raised by patients to the hissing sound referred to, and have on the other hand always endeavoured to keep the apparatus as simple and free from complications as possible.

3. Mouth-props are used to place between the teeth during the inhalation, and so avoid the loss of time which would ensue if it were necessary to overcome the spasm of the muscles of the jaw after removal of the face-piece. They should be provided and fitted into place by the operator, prior to the commencement of the inhalation, but the administrator is frequently called upon to do this; he should therefore be provided with one or two pairs.

Figs. 15 to 18 represent various forms of these props, which are too simple to require any explanation.

In selecting a mouth-prop the following points should be attended to :—

  a. It should be made of hard material, not likely to split or chip, so that it may be washed and scrubbed frequently. Pads of rubber or some non-absorbent substance, may be fitted to the dental surfaces.

  b. It should be as small as is compatible with strength, or it will impede rather than assist the operator.

  c. It should be as simple as possible, as joints are likely to give way or fail.

  d. A stout piece of catgut, silk, or string, eight or ten

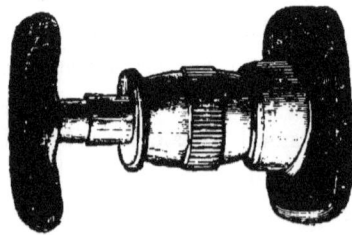

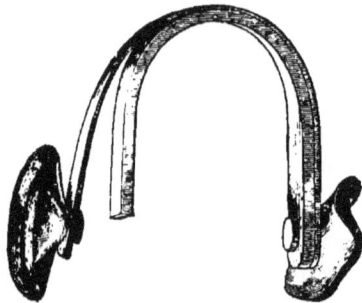

MOUTH PROPS.

FIG. 15.—Buck's.　　FIG. 16.—Spring and Screw.　　FIG. 17.—Hutchinson's.　　FIG. 18.—Simple.

inches long, should be tied firmly round the stem, and attached to another prop, and this string should be renewed with every fresh patient.

4. Gags and mouth-openers (Figs. 19 and 20) are of some service should the prop slip during the extraction, or when, as sometimes occurs, it is impossible to adjust them before

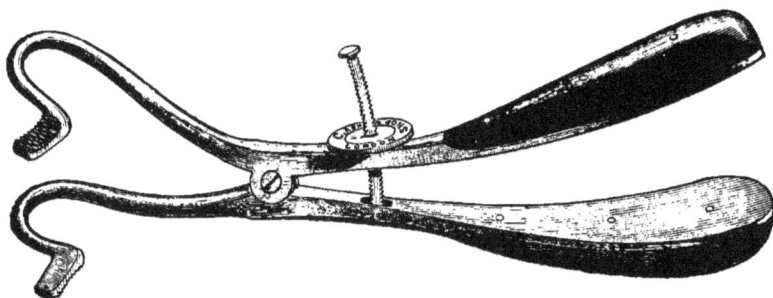

Fig. 19.—Coleman's gag.

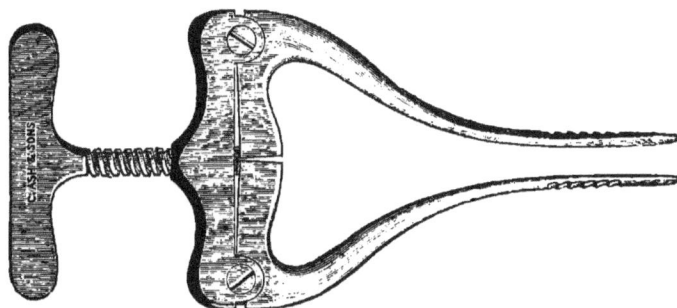

Fig. 20.—Mouth-opener.

commencing the inhalation, e.g., in children or in certain morbid conditions in adults.

5. Tongue forceps (Fig. 21), to seize and draw forward the tongue in the event of dangerous symptoms intervening.

6. Throat forceps (Fig. 22), for the extraction of teeth or other foreign bodies that may accidentally slip into the larynx.

FIG. 21.—Tongue Forceps.   FIG. 22.—Throat Forceps.

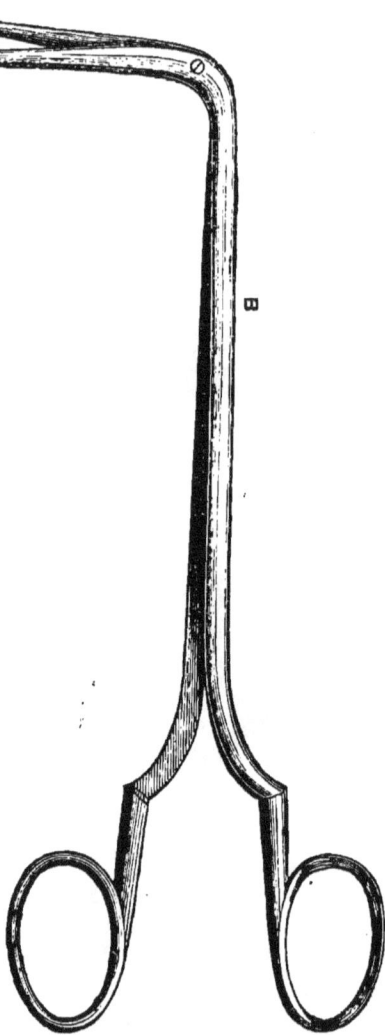

7. A few sponges, loose, and fixed firmly to handles, to wipe out the pharynx should accumulated blood and mucous threaten asphyxia.

8. Nitrite of amyl in capsules, and a little strong ammonia, for use in the event of syncope.

9. Tracheotomy instruments should also be at hand.

I need hardly insist upon the importance of keeping all these instruments scrupulously clean. The mouth-props and gags especially, must be well washed and scrubbed, not simply dipped in water, each time after they are used.

CHAPTER V.

Preparation of Patient and Apparatus.

Before proceeding to explain in detail the methods and phenomena of actual administration, a few preliminary remarks are necessary.

In the first place, it must be laid down as a rule which admits of no exception, that *the administration must occupy the sole and undivided attention of one individual*, or, in other words, must never be undertaken by the person who is about to perform the operation. Not to mention the obvious and great risks which single-handed administration involves, and the grave responsibility incurred in the event of any accident, it must be recollected that the production of anæsthesia by nitrous oxide is a question of seconds, and that its duration is equally brief, so that even the trifling loss of time consequent upon changing the face-piece for the instrument is of importance, however skilfully the change may be effected; not only so, but in attempting to combine the two functions, one of two things almost invariably happens, viz.: in the anxiety to obtain as much time as possible, the patient receives a little more gas than absolutely necessary, and there may be in consequence, to say the least, unpleasant after-effects; or the reverse may occur, and the anxiety to operate predominate, too little gas is then given, and the patient becomes semi-conscious before the operation is completed, and it has been proved over and over again, that

susceptibility to shock and consequent tendency to syncope, is
much more marked in this semi-anæsthetised state, than even
if no anæsthetic at all had been given; in neither case does
the ambitious individual obtain much credit, and runs considerably more risk than any wise man would willingly incur;
the gas, too, is brought into disrepute, and shares the blame
which rightly should belong solely to the mal-administration.
Not only so, but in one case of fatal syncope which has been
reported, the operator was so intent upon his work, that he did
not perceive, until the extraction was completed, that the
patient had fainted, and was in fact dead before the last
tooth was drawn; had he been aware of the condition some
seconds earlier, it is more than probable that life might have
been restored.   Further, in the case of females, single-handed
administration may form the basis of cruel fabrications and
groundless charges, which, however false, are nevertheless a
source of serious annoyance and worry.   It may not perhaps
be out of place to mention, that in Ohio a law exists, render-
ing the presence of a third person compulsory, in all cases in
which an anæsthetic is administered, for whatever purpose.

Some remarks of an eminent living authority on anæsthetics, may serve to emphasise the importance of this subject.

Mr. Braine observes :* " Sensual emotions are not un-
frequently excited in both sexes. . . . An unmarried
hysterical girl certainly gave evidence, by her movements, that
she was quite aware of one of the duties of married life ; and,
moreover, in this case, the idea was still present when she was
able to speak, for she addressed the administrator in terms
far fonder than the occasion warranted, while another girl,
who had behaved in a similar manner, said, ' I hope I have not
said anything naughty.'   Both of these cases brought forcibly
to one's recollection many trumped up cases of felonious

* " British Medical Journal," January 23rd, 1839.

assault; and how extremely unadvisable it is to have recourse to anæsthesia without a third person being in the room."

To the rule above enunciated a corollary may, with advantage, be added, viz.: that the administrator should always be a qualified man, capable, by training and experience, of dealing promptly and successfully with such emergencies as may arise. Happily, cases of death while under the influence of gas are rare, but fatal cases have occurred and may occur again at any time; it is better therefore always to be prepared for the worst, and it is needless to point out, how great an influence the evidence of a qualified man would have upon the minds of a jury, in the event of any judicial investigation being made; and although actual death is of comparatively rare occurrence, difficulties and dangers have not infrequently arisen, in which the saving of the patient can be distinctly attributed to the action of the medical man in attendance.

The next question which arises is, "Are there any cases in which the inhalation of the gas should be absolutely forbidden?" This question is usually answered either by an unqualified negative, or by the assertion, that if the patients are fit subjects for operation, they are *ipso facto* capable of inhaling the gas with impunity. But inasmuch as the safety and future well-being of the individual is concerned, we cannot dismiss this subject thus cavalierly, or by the use of a mere dogmatism. M. Laffont (*op. cit.*) quotes cases in which its administration appears, at least, to have been followed by premature confinement, menstrual disturbance, and epileptiform seizures in the previously healthy, and by exacerbations of cardiac disease, and of glycosuria, in those previously subject to such complaints, and advises that it should be forbidden in these cases, and should not, under any circumstances, be administered, without the consent and assistance of a medical man. These gloomy prognostications have not been verified, and are not generally accepted, but a careful consideration of what we know concerning its physiological action and pathological effects (Chapter II), cannot but lead us to the conclusion, that elements of danger *must* exist, where the physiological processes are so profoundly affected. Against this pessimistic view, we must place the results of actual clinical experience; the number of successful administrations may now be reckoned by hundreds of thousands, or even millions, but we must recollect that, although the administration of the gas, has been attended with remarkably few fatal accidents, numerous cases are upon record, in which the life of the patient has been placed in jeopardy, or has only been saved by the skill and promptitude of the attendants. Not only, therefore, from a theoretical, but also from a practical point of view, I think that the direct negative must be modified, and that in reply to the question enunciated above, we must say, that although the gas has been safely inhaled by patients suffering from almost every possible morbid condition, its administration when either the circulatory or respiratory systems are known or suspected to be affected, entails considerable risk, calls for special skill and care, and should *never*, under these circumstances, be attempted, except in the presence of a medical man. The "assertion" alluded to, savours of an attempt to transfer to the shoulders of the operator, a responsibility which rightly belongs to, and should be fully accepted by, the administrator; as in the case of many other trite sayings, the result of their application is in a large majority of cases practically true, and when at fault the failure is said to "prove" (not probe) the rule, but in all cases their logic is extremely bad, and consequently open to misinterpretation and abuse.

E 2

Of course the operator is, in many cases, much more likely
to know of the peculiarities of constitution, and physical
condition of the patient, than is the administrator, who,
probably, sees the individual in question for the first time,
immediately prior to the operation. But, on the other hand,
it is the undoubted duty of the operator to impart his knowledge to the administrator, and to allow him, and him alone,
to judge, as to the fitness of the patient for being anæsthetised,
as to the anæsthetic to be employed, and as to the exact
method of its administration.

PREPARATION OF THE PATIENT.

Unlike ether, chloroform, and their allies, little or no
previous preparation in the shape of fasting, is either necessary or advisable ; the gas should not be inhaled immediately
after a full meal, in case a tendency to vomit be thus
encouraged, due rather to the blood swallowed than to the
effects of the gas ; on the other hand, prolonged and forced
abstinence may increase the liability to syncope, and to disagreeable after effects.

Before the patient is introduced into the operating room,
the following few simple preparations should be made.
In young children, and hysterical females, the bladder and
rectum should, if possible, be evacuated, as in this class of
patients involuntary micturition and defæcation sometimes
occurs under the gas, a proceeding which is as embarrassing
and disagreeable to the patient as it is to the operator.
The upper buttons of the coat or dress may be loosened,
and the collar and brooch removed, especially if the latter
are at all tight, or likely to get in the way of the operator ; if
spectacles are worn, they should be taken off, and eye-glasses
removed from the eyes. Artificial teeth should be taken out,

in case they become loosened, when they would tend to fall into the larynx, or impede the manipulations of the operator. Gloves should be removed from the hands, as the latter form important points of observation for the administrator, as will appear in the next chapter. If the patient complains of faintness, a little weak brandy and water, or sal volatile, can be given with advantage.

The patient is now introduced into the operating room, and will probably for the first time, make the acquaintance of the administrator, and it is of importance, for the comfort of all concerned, that the latter should, as far as the short time at his disposal will permit, gain, or attempt to gain, the confidence and trust of the former. It is of course quite impossible to lay down rules of any description, children especially require considerable tact and care, but even with them, and certainly with most adults, the *suaviter in modo* must be judiciously combined with, but must never entirely displace, the *fortiter in re*. Above all, and before all, the administrator must be ever mindful of the importance of the trust committed to his charge, and must recollect, and fully allow for, the nervous condition in which the patients are generally found. He should also accustom himself to " take stock," so to speak, of the patient ; the first glance of the well-trained eye will convey a large amount of useful information, and may suggest many little modifications in procedure, or lead us to anticipate a troublesome administration or the reverse. He can thus be both forewarned and forearmed against the bronchitic, the alcoholic, the nervous, the anæmic, the plethoric, and many phases of hysteria, &c. He should further inform himself of the exact character of the operation about to be performed, and, in fact, should make himself as thoroughly acquainted with the probable movements of the operator as possible.

The patient's friends should be distinctly told, that they are themselves much more likely to be disturbed by the sights and sounds of the operation, than is the patient himself; but, on the other hand, it is as well not to insist upon excluding them altogether, yielding as gracefully as possible in deference to the express wish of the patient. Do not, however, allow more than one friend to be present, who must undertake, not in any way to interfere with the proceedings, to do exactly as the administrator or operator may wish, and must be placed in such a position as not to be able to see the face of the patient, who, as will appear, presents, when fully under the influence of gas, an appearance far from pleasing to the uninitiated.

Another question which arises at this point is, Should the patient be examined in any way, and if so, how? Mr. Braine, and other authorities on the subject, advise that a superficial examination should always be made, such as feeling the pulse, and ascertaining that the chest-walls expand fairly well and equally (by simply placing the hands on the chest, and directing the patient to take one or two deep inspirations); but this is with a view to satisfy the patient, and to put one's-self on the right side, in the event of accident and consequent awkward questions that might be asked, rather than to obtain any very precise information for one's own guidance. Sir Joseph Lister, alluding more particularly to chloroform, but the opinion is if anything more applicable to nitrous oxide, maintains that even such examination is quite unnecessary, and, in fact, tends to alarm the patients and raise suspicions in their minds which were previously non-existent. I think, then, that it may be laid down as a pretty safe rule, that if the patients or friends suggest any doubt as to the former's fitness for being placed under the influence of the gas, or express, even indirectly, by word or sign, (such as holding out the hand for the pulse to be felt), a wish for an examination, it is as well to adopt the superficial method suggested above, rather than attempt to overcome their scruples by argument, while, if definite lesions are known or suspected to exist, a more thorough and complete investigation should, of course, be made. Under other circumstances the less the equanimity of the patient is disturbed the better.

ARRANGEMENT OF APPARATUS.

All the preliminary arrangements of the apparatus and instruments, should be made before the admission of the patient into the operating room, so that no unnecessary delay is experienced in commencing the actual inhalation, and the patient's nervousness is not increased by dwelling upon the sight of the apparatus.

If working from a gasometer (Fig. 23), fill the inner reservoir (I) from the bottle (3), fit the face-piece (D) tightly to the two-way tube (C), previously attached to the tube (B), assure yourself that the gas itself is pure, that the taps and valves are working properly, and that there is no obvious leakage; if doubtful, turn the taps and take one or two inspirations yourself through the face-piece, and in any case allow a little gas to flow through the tube in order to expel the air. If a supplemental bag is to be used, this must be attached to the face-piece as at (E), and if at all worn should be first examined for cracks or leakages, by distention with air, if necessary. The figure on the following page represents the apparatus fitted for inhalation from a gasometer.

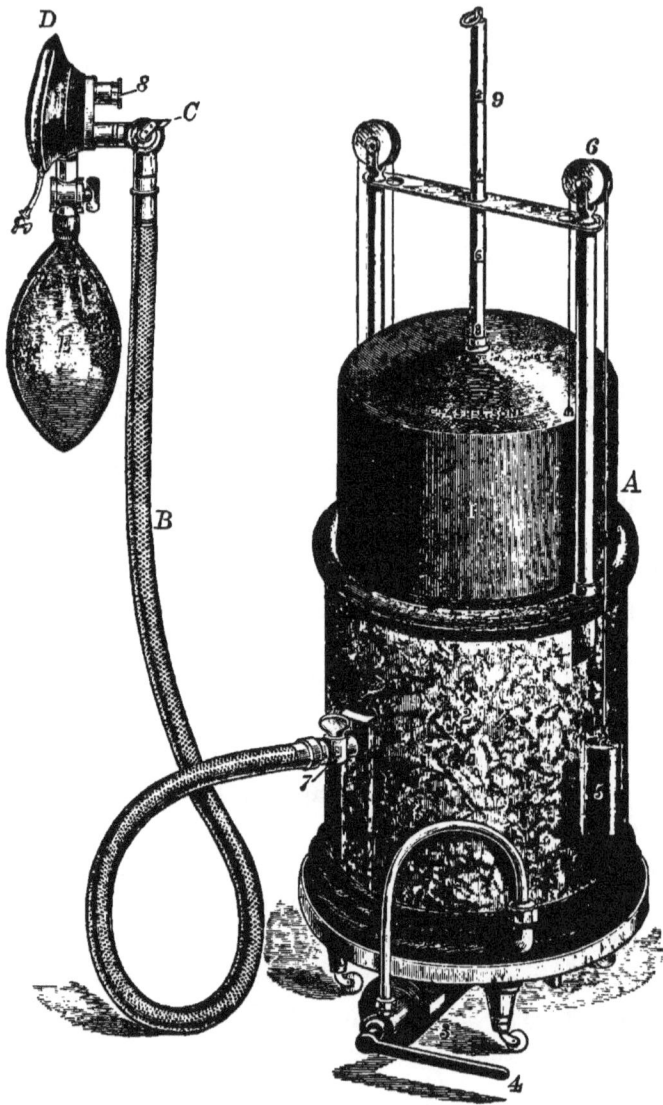

FIG. 23.—Apparatus arranged for inhalation from a gasometer.

If the direct method is to be employed, fit the apparatus as in Fig. 24, in which the position of the horizontal twin bottles with foot pedal, stout conducting tube, reservoir bag

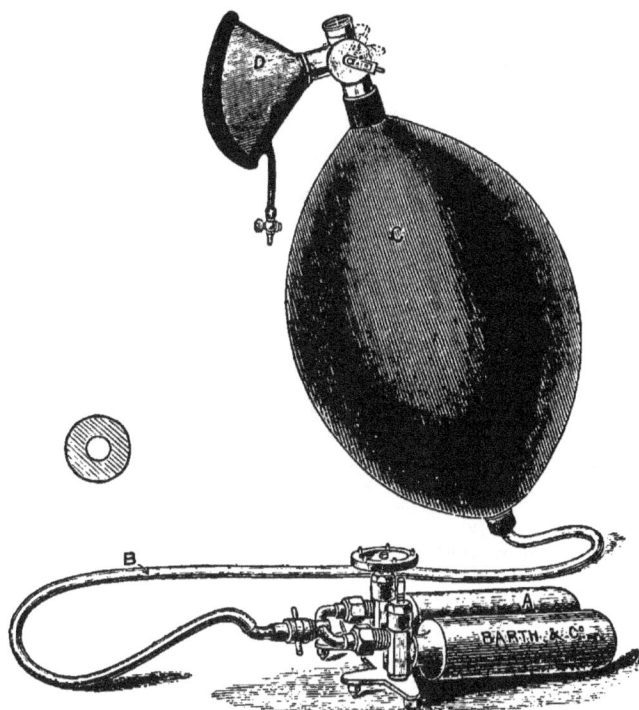

FIG. 24.—Apparatus arranged for direct administration. A, gas bottles ; B, conducting pipe, a section of which is represented above ; C, reservoir bag ; D, face-piece.

brought close up to the face-piece, three-way tube, and conical face-piece are shown.

Fig. 25 represents an apparatus with a wide tube interposed between the reservoir bag and the face-piece, the horizontal bottle turned by the hand of an assistant being contained in a portable box or case.

If either of these arrangements is used, assure yourself

FIG. 25.—Apparatus arranged for administration by the direct method.

that there is plenty of gas in the bottle you propose to work from, or transfer the pedal to a full bottle. See that the nuts and screws connecting the gas bottle with the con- ducting tube are air-tight, and that the tube itself has no obvious leak or hole. If the reservoir bag or pad of the face- piece is at all hard or stiff from disuse or cold, warm it at the fire before attempting to manipulate it, or it will crack. Be sure that the stopcock of the three-way tube is working properly and quietly, that the valves of the face-piece are acting efficiently, and that the oral pad is fully distended, and air-tight. The stopcock of the three-way tube being turned off, allow a little gas to flow into and partly distend the reservoir bag, then turn on the stopcock, and press out the mixture of gas and air in the bag by rolling it up tightly in the hand, and, before releasing it, turn off the stopcock again ; by this means the air will be expelled from the conducting apparatus and its connections, and the gas may then be allowed to flow gently into the bag until the latter is quite full, and all will be ready for the reception of the patient.

This latter account of the preparation of the apparatus takes much longer to write or read than it does in actual practice : once fitted up, the apparatus, whether gasometer or direct, can always be kept ready for use, and all that is then necessary is to expel the air from the tubes and fill the reservoir bag.

If a rarefier or silencer (Figs. 13 and 14) is used, it is attached directly to the gas bottle, and the former must, of course, be filled with hot water before the gas is turned on.

The mouth-gag (Fig. 19) and tongue forceps (Fig. 21), especially the former, should be placed in a handy position, (by preference in an empty pocket of the administrator), and the exact position of the instruments of emergency—*e.g.*, capsules of nitrite of amyl, throat forceps, &c.—well ascertained.

CHAPTER VI.

ADMINISTRATION.

WE will first consider the method of administration and phenomena observed, in an ordinary typical case, the apparatus represented in Fig. 24 being used, and the patient being an adult female, and will reserve for future chapters the account of such difficulties and dangers as may arise, the consideration of administration in special cases, and of any possible variations in the symptoms that may call for remark.

For purposes of description this portion of our subject may be discussed under three heads, viz. :—

I. Preliminary,

II. Actual Inhalation,

III. Recovery,

and each of these must further be considered both in relation to the patient and to the administrator.

PRELIMINARY.

Commencing with the patient taking up her position, and terminating with the application of the face-piece.

The Patient.—The preparations according to the suggestions laid down on page 52 should have been made before entering the operating room, so that as little time as possible is lost in commencing the actual administration; but many of the details—such, for instance, as removal of artificial teeth, taking off gloves, loosening buttons, &c.—may be, and are most frequently, left until the last moment.

One of the great advantages possessed by nitrous oxide over other anæsthetics is, that it may be administered with equal facility and safety in almost any position; that usually adopted by dentists is the one to which the following account is more particularly applicable. The patient is seated comfortably, and without restraint, well back in the operating chair; the shoulders should be supported, and the head nearly in a line with the body, or but slightly tilted backwards. The hands should be lightly crossed in the lap, not clasping either the chair or one another.

She is now directed to open her mouth as widely as possible, and one or other of the props, figured on page 44, or some similar contrivance, is introduced between the teeth, and at the same time, a glance is given round the mouth to assure one's-self that no artificial teeth, (the existence of which some people are loth to acknowledge), remain in position; in placing this prop, the following points are of some importance and should be attended to :—

(a) Preference should be given to lateral over anterior positions.

(b) The prop should be inserted between teeth, if possible, not on the bare gum.

(c) Always choose sound teeth, and avoid single teeth unless perfectly sound and firm.

(d) Fix the prop firmly and flatly on the teeth, not at an angle, pulling lightly on the string to assure yourself that it will not slip.

The next step is to direct the patient to breathe deeply and regularly, filling the lungs well at each inspiration; do not content yourself with simply *telling* her how to breathe, but show her, fitting the apparatus to your own face for a few seconds by way of demonstration. The face-piece, arranged for the admission of air only, is then taken into the hand, (it matters not which), and adjusted carefully over the nose and mouth of the patient, at first lightly in order to ascertain that it fits accurately, and that the regular breathing is maintained, it is then pressed firmly into position, and, selecting an interval between a movement of expiration and the following inspiration, the stopcock is turned on to admit a full flow of gas.

All this takes some time to describe, but the whole process in practice should not exceed a few seconds, and I must insist most strongly upon its importance, and more especially upon the accurate fitting of the face-piece, as upon this latter point in particular depends much of the success of the administration ; a little trouble and care at this stage will be amply repaid to the patient by a sense of security and comfort, and to the administrator by the rapidity and normal development of the phenomena.

The Administrator.—The best position for the administrator to assume is behind the chair, leaning over the patient from above ; he is then able to see both sides of the face, has greater command over the face-piece, can detect any leakage beneath the pad, and can apply the necessary pressure more firmly and equally. Should the operator wish to work from behind, the administrator can, towards the end of the inhalation, gently shift his position, carefully maintaining firm hold of the face-piece ; or the gas can be administered from the side from the beginning, the side chosen being of course opposite to the operator (*i.e.*, usually the left); it is as well, too, to pass one of the fingers of the hand holding the face-piece beneath the chin, and, by gentle pressure, assist in maintaining the position of the mouth-prop.

The gas bottles should be arranged at the feet of the administrator, who stands with his foot lightly pressing upon the pedal; the tube connecting the gas bottle with the reservoir bag, being allowed to fall so as not to be likely to catch in anything when the face-piece is suddenly removed. The hand not employed in holding the face-piece should softly feel for the superficial temporal artery, which will be usually found about an inch, or an inch and a half, above and external to the outer canthus of the eye, though subject to much variation in position and size in various individuals: a mental note should be made of its exact position, and of the character of the pulse it contains, or, if the administration be conducted from the side, the state of the pulse at the wrist may be determined.

ACTUAL INHALATION.

Commencing with the application, and terminating with the removal of the face-piece.

The Patient.—As has been suggested in the chapter on physiology, the production of anæsthesia is by no means a simple process, but the resultant or average, so to speak, of at least two very distinct and opposite stages, viz. :

(*a*) A pre-narcotic stage in which there is more or less excitement, both mental and physical, and

(*b*) A narcotic stage gradually extending and deepening.

From a clinical standpoint, it is often difficult to determine the symptoms respectively referable to these stages. The pre-narcotic stage especially, may appear to be exceedingly brief or even, as far as objective signs are concerned, altogether absent, but it is seldom that subjective symptoms cannot be obtained, when the patient is questioned on return to consciousness. The course of events in a case such as we are now considering (*i.e.*, a typical case) is usually as follows :—

The first few inhalations, produce in the patient subjective feelings of fulness and throbbing in the head, followed by tinnitus and more or less mental exhilaration; objectively, the face becomes pale, or rather of a leaden hue. Exaggerations and perversions of special sense, *e.g.*, hearing, sight, &c., not unlike those occurring in commencing sleep, are now developed, and should warn us against disturbing the absolute silence, so essential to an operating room. At the same time, too, it is probable that the dreams, more or less pleasurable and associated with sense of rapid, noisy movement, *e.g.*, railway travelling, occur, though of course it is quite impossible to fix the exact period of their origin.

The ashy hue of the face deepens to one of marked and gradually increasing lividity, the mucous membranes of the lips becoming of a blue-black colour; this lividity is not of necessity a sign of insensibility, nor of venous congestion, but of vascular dilatation and darkening of the blood, it need not therefore cause the least anxiety; the eyes soon become fixed, the conjunctivæ slightly congested, and often, even at this early stage, wholly or partly insensitive to touch; pinching the skin will now produce no signs of pain.

Continuing the inhalation, the breathing becomes slower and usually snoring in character, slight tremors in the extensor tendons of the thumbs will soon be observed, and at or about the same time the conjunctivæ will be found to be quite insensible, the eyeballs oscillate, the eyelids twitch in a manner suggestive of epilepsy, and the pulse, which should always be watched during this part of the inhalation, becomes appreciably slower.

The first appearance of tremor or twitching is a sign that

a sufficient degree of anæsthesia has been produced for the commencement of the majority of small operations, or the extraction of one or two loosely fixed teeth ; but if the operation is likely to last a little longer, we may with safety allow a few more inspirations under increased vigilance. The snoring respirations will then become truly stertorous, and show an inclination to intermit, the convulsive twitchings will extend to the arms and legs, amounting almost to jactitations, and a condition of tonic spasm set in. The pupil is commencing to, or has actually become, widely dilated, the pulse distinctly intermittent.

Beyond this it is dangerous to proceed, and for my own part, I seldom think it necessary or advisable to proceed beyond the development of true stertor; the very trifling prolongation of the anæsthesia then obtained, is more than counteracted by the trouble in restraining the movements of a spasmodic patient. I would also warn my readers against relying upon the dilatation of the pupil as a sign of the completion of anæsthesia; although it very frequently occurs either immediately before or after removal of the face-piece, I am personally inclined to look upon it, at best, as an accidental phenomenon, and inasmuch as it may suggest the approach of syncope, rather as a warning to keep the finger on the pulse and watch the patient more narrowly.

Although I have alluded to the development of the subsultus as the best, and most constant sign of the completion of anæsthesia, it must be fully understood, that no *absolute* rule can be laid down as to when to withdraw the gas. Each case must be judged upon its own merits, and due regard had to other symptoms, such as stertor, intermittent breathing, state of the pulse, &c.

But while on the one hand one must be careful not to overdose the patient, on the other, it is of almost equal

F

importance not to permit the operation to be commenced, before the patient has completely passed under the influence of the gas. The majority, if not all the accidents that have occurred during inhalation, have been due to the intentional or accidental neglect of this precaution.

As to the length of time required to produce the full physiological effects—*i.e.*, stertor, &c.—this of course varies greatly with the individual. The Odontological Committee places the average for men, women, and children at 73·3 seconds. Dr. Hewitt,* as a result of six observations on five different days upon the same female, puts it at 68·6 seconds; in eleven cases in which I have myself noted the length of time taken in adults of both sexes, I obtained an average of 67·8 seconds. Giving an average all round of 69·9 seconds for each case.

The amount of gas required under similar circumstances is likewise liable to great variation, and not only according to the method of administration, and the idiosyncrasies of the patients, but apparently also, though of course to a less extent, with the temperature of the room and with the barometric pressure. Three hundred (300) observations by Dr. Hewitt at the National Dental Hospital gave an average of 6·9 gallons per head, but this is I think much too high; as the result of twenty-nine observations upon 309 patients at the same institution, I have myself obtained an average of 3·5 gallons per head, the average age of the patients being 24·8 years. The point is not one of any very great importance, but the question is often raised by students and others, and a correct estimate would also be of value to the practical anæsthetist in assisting him to detect, or suspect, leakages from his gas bottles, &c. Of course, the amount of gas required to anæsthetise a series of, say, 10 cases (as in hospital

* "Journal of British Dental Association," June 15th, 1886.

work), is, for many reasons, likely to be much less than that which would be necessary for the same number separately, or in batches of two or three (as in private) ; but after making every allowance in this direction, I think we should be justified in suspecting that something was wrong, if a 50-gallon bottle of the liquid did not serve for about twelve separate administrations.

The Administrator.—Immediately after the first expiration, the valve of the gas bottle should be very slightly opened by rotating the heel of the foot resting upon the pedal to the right, so as to allow the gas to flow very gently, and with as little noise as possible into the reservoir bag ; after five or six more respirations the valve may gradually be turned on more fully, until the gas is flowing pretty freely ; after another six or eight respirations have taken place the reservoir may be converted into a supplemental bag, by turning the stopcock of the three-way tube to the third position (see page 42); the bag will then rapidly become distended and the gas must be shut off.

The operator should now be warned by a sign that the patient is nearly ready, for a very few respirations, in and out of the bag, will suffice to complete the anæsthesia; the face-piece is then quickly removed and may either be allowed to drop into the lap of the patient, or on to the floor, or may be placed on an adjacent table, in any case in some place where it will be well out of the way of both administrator and operator.

RECOVERY.

From the removal of the face-piece, to the complete return of consciousness.

The Patient.—If the patient has passed thoroughly under the influence of the gas, she presents the following appearance upon removal of the face-piece. The lips and adjacent mucous surfaces are of a blue-black colour, the skin of a livid hue, or if thin, distinctly blue ; at the same time the pulse is slower and feebler than usual, and inclined to intermit or be irregular in rhythm. The respirations are noisy, stertorous, and spasmodic in character. The eyes projecting and staring, the conjunctivæ congested, and insensitive to touch, the pupils in variable conditions of dilatation. The muscular system is profoundly affected, and spasm of various kinds, invariably present, may be simply of a clonic character and confined to twitchings of the hands and eyelids, but often, if not always, there is more or less tonic spasm and rigidity of the muscles. Added to this the fact, that the mouth is held widely open by the prop or gag, and it will be readily understood why we should be anxious to place the patient out of sight of the friends ; to the initiated, however, this condition of affairs need cause not the slightest alarm. The narcotic quickly gives place to the stage of excitement; the pulse returns to its normal rhythm with almost the first aërial inspiration, the lips and skin regain their usual hue of health or become even brighter, and almost at the same time the stertor disappears, and is replaced by quick, shallow, or even panting respirations. The congestion of the conjunctivæ may remain a little longer, but its insensibility soon disappears. The convulsive twitching, or clonic spasm becomes merged into tonic spasm of short duration—in fact, the recovery is, as a rule, as complete as it is rapid, and beyond a dazed feeling no ill effects are complained of.

As soon as possible after the completion of the operation, the head and shoulders of the patient should be drawn well forward, to prevent the blood flowing back into the larynx but beyond this, provided the lips are resuming their natural colour, and the pulse and respirations are becoming gradually normal, no violent efforts to rouse the patient should be made, and the over-anxiety and pressing inquiries of friends are especially to be deprecated. When the patient herself gives signs of returning consciousness, she should be encouraged to wash out the mouth, or perform any little voluntary acts, such as taking a deep breath, holding the glass or spittoon, &c.; this draws her attention away from herself, and helps to prevent hysterical attacks.

After washing out the mouth thoroughly, the bleeding having stopped, the patient should be allowed to lean back comfortably in the chair for five or ten minutes before attempting to rise, she may then be transferred to another room, and may safely be allowed to leave the house in the course of another ten or twenty minutes, and may be assured that no ill effects are likely to follow, and that no special care or regulation as to diet is at all necessary.

The Administrator.—The duties of the administrator do not cease with the removal of the face-piece; he must then watch the face of the patient for signs of returning consciousness, or the reverse, and give the operator warning, when the anæsthesia has passed off and the patient become sensible of pain.

It is of importance, too, that full use should be made of the anæsthesia induced, and it is therefore the duty of the anæsthetist, when the safety of the patient is assured, to watch the mouth-prop, and if it shows signs of slipping, he must rapidly insert the gag (Fig. 19) between the jaws, without waiting for the operator to ask him to do so, taking care to choose the side opposite to that upon which the operator is working: he must also be ready to restrain any involuntary reflex movements on the part of the patient that might interfere with the manipulations. If the mouth-props do not slip, or are not knocked out in the course of the extraction, it is as well not to remove them until complete consciousness is established, or the patient may imagine, that another tooth is being extracted, and complain accordingly.

The time which elapses between the removal of the face-piece and the return of consciousness has been variously estimated; on the average it would appear to be about 36 seconds (Odontological Committee 24·6 seconds, Dr. Hewitt (*op. cit.*) 47·5 seconds). This, however, must not be confounded with the total duration of nitrous oxide anæsthesia, which is probably much longer. It must be recollected, that our object in administering the gas according to the method above described, is to obtain the longest anæsthesia possible in one application of the face-piece, this is synonymous with the deepest possible narcotism : simple sensation is probably abolished quite early, and when the inhalation may be continued at the same time as the operation is being performed, (*i.e.*, not upon the face), the actual anæsthetic condition may be considered as lasting for a minute and a half to two minutes.

CHAPTER VII.

VARIATIONS AND MINOR DIFFICULTIES OF ADMINISTRATION.

IN the preceding chapter I have described, as succinctly as is compatible with clearness, the phenomena, and methods of administration, in what may be termed *an ordinary typical case,* but, unfortunately, a very large proportion of the cases with which the anæsthetist will be called upon to deal, are by no means "typical." The phenomena of nitrous oxide administration, from first to last, vary considerably, not only in themselves, but also in the sequence in which they are developed, and we must, therefore, consider the subject of actual administration a little more in detail, and by the light of the physiological facts with which we are already acquainted (Chapter II), and must discuss certain departures from the ordinary routine, both of manipulations and phenomena.

VARIATIONS IN PROCEDURE.

If any other form of apparatus is used, than that to which the above descriptions of the manipulations during the first two stages more particularly apply, it will not be difficult to make such alterations as are necessary, bearing in mind the principles involved. Thus, if the arrangement represented in Fig. 23 be used, the gasometer should be placed behind, and a little to the left of the chair, or it may be placed directly in front and to the left of the patient, on a shelf in the corner of the room for instance. The stopcock of the gasometer having been turned on, that attached to the face-piece is opened, according to the directions already given, and if it is desired to use the supplemental bag, the stopcock of the latter is turned after ten or twelve respirations, and the finger is placed upon the expiratory valve of the face-piece (Fig. 23–8), in order that the bag may become distended.

In adjusting the face-piece considerable care is necessary, in order to obtain an accurate adaptation to the irregular outline of the nose and mouth; it is as well, therefore, after applying it, to run the finger round the edge, to ascertain that no very obvious interval exists between the face itself and the air pad. If, after inhalation has commenced, a leakage is suspected, and it does not appear of sufficient size, to warrant the removal of the face-piece and re-administration of the gas, efforts should be made, if in the face-piece, to control it with one of the fingers, and a free flow of gas from the reservoir bag should be maintained, by lightly pressing upon the bag itself with the disengaged hand, keeping up a full supply from the bottle at the same time. This, by the way, is another argument in favour of having the reservoir bag close to the face-piece—i.e., in the event of a leakage it is more easily accessible.

Turning the stopcock of the face-piece, in order to admit the gas, should be done rapidly, so that little or no air is drawn in, and without any " click " or " snap," which, in the nervous condition of the patient, may become exaggerated, and lead her to conjure up imaginary evils in the shape of " something having given way," or the thumb may be placed over the air-hole, while the stopcock is turned gradually on with the index finger.

Beards and whiskers are a source of trouble to the

anæsthetist, inasmuch as the hair affords ready means for leakage beneath the pad of the face-piece. It is as well, before the administration commences, to mat the hair together with vaseline or some stiff pomade, common soap answers admirably for this purpose, but is disagreeable and difficult to remove; in the absence of either, wet the beard well with water. It is in the avoidance of this trouble, that the American forms of mouth-piece have their special advantage, but no very extended trial has been given them in this country.

VARIATIONS IN PHENOMENA.

Breathing.—The first trouble we are likely to meet with, is that due to the nervous condition of the patient, who, directly attention is drawn to the manner and rate of breathing, or directly the face-piece is applied, either insists upon holding the breath for very considerable periods, or breathes in a very shallow and perfunctory manner, and, maybe, struggles, and endeavours to remove the face-piece; if a little precept and example be not sufficient to overcome this difficulty, remove the face-piece, cover the eyes lightly with the hand, or a folded towel, and gently restraining excessive violence, and keeping the room quite quiet, allow a few ordinary respirations to take place; the patient usually becomes composed in the course of a very short time, and the inhalation may be proceeded with as if nothing had happened.. In some cases, however, notably in children, this has little or no effect; it is then probably better to force the inhalation in spite of the struggles, taking care that the patient neither injures herself nor breaks any adjacent objects, and in such cases, especially, be sure that a free and full supply of gas is obtain-

able, as the inspirations are then usually very deep, and
rapidly exhaust the reservoir bag, not only so, but the
struggles have been known to be due to actual deprivation
of gas, consequent upon defective valves.

Cough.—Should the first entrance of the gas cause coughing and struggling, it is frequently a sign that some impurity
is present, and if this cough tends to become intensified as
the inhalation proceeds, we must remove the face-piece, taste
the gas ourselves, and if our suspicions are confirmed, we must
obtain the gas from another source.   Cough induced at a
later stage of the inhalation, when the anæsthesia is nearly
complete, is more often than not due to throat irritation, and
is then frequently the effect of a long and relaxed uvula upon
the fauces, base of the tongue, &c., and this is one reason why
the head should not be tilted too far back.

Excitement.—The pre-narcotic, or stage of excitement may,
instead of being hardly appreciable, become unpleasantly
obtrusive, and in hysterical girls and women, every variety
of antic may at times be indulged in, of which singing and
kicking are the most common, and hallucinations of an erotic
nature not infrequent.   The indication in these cases is
undoubtedly, to force the inhalation as much as possible, by
pressure on the bag, and a full supply of gas; do not attempt
to restrain, or argue with the patient, so long as she does
not injure herself or others, and as a rule a very few deep
inspirations suffice to overcome the trouble.   Co-ordinated
movements of a rhythmic character, such as beating time to
music, swinging the legs, &c., often show themselves during
the first few inhalations of the gas, but disappear after three
or four inspirations; they should not be interfered with,
beyond protecting the limbs from injury.

Occasionally hysterical girls, hitherto quiet, commence
to scream and kick just as they are passing fully under the

influence of the gas, and we are thinking of removing the face-piece. They should not be restrained in any way, and the movements will usually subside if the gas is pushed. Such movements are usually quite reflex in character, and even if they do not subside, we need not, on their account alone, hesitate to commence the operation, as the actual anæsthesia is not interfered with.

During the stage of recovery, too, the period of excitement is often very marked, especially in females : hallucinations, with a desire to go somewhere or do something, are very common ; there may be also more or less violent screaming, beating of the feet, jactitations, &c., followed by hysterical crying. The best antidote to this condition, is full exposure to the fresh air obtained by opening an adjacent window, but where this is impossible, the patient should be prevented from injuring herself or others, and, as soon as the first glimmer of consciousness returns, should be induced to perform one or other of the little voluntary acts suggested on page 69, or the hand or a napkin may be gently held across the eyes until the attack has passed off.

The admission of air into the apparatus by a small leakage, or beneath the face-piece, is not usually associated with actual excitement, as is sometimes supposed, but the production of anæsthesia is then much prolonged, or rendered absolutely impossible ; this is because the amount of air actually admitted is small, but if large, excitement is likely to occur.

Lividity.—As a rule lividity is a very constant and very early phenomenon ; its absence or any delay in its development must make us at once suspicious of the admission of air.

In children and people with delicate skins it is usually marked, the lips and even the cheeks becoming of a distinctly blue colour. In elderly people, or in others where the blood is already imperfectly oxygenated, the mucous membranes may become nearly black.

Ophthalmic Changes.—The condition of the eye is very variable, the conjunctiva may remain sensitive for a very considerable time, and, in fact, its reflex response to mechanical irritation, may not be completely abolished until quite the end of the administration, when other symptoms warn us that it is time to discontinue the gas; in elderly people the congestion is frequently very marked.

The pupil, especially in hysterical and anæmic women, may become widely dilated at quite an early stage of the inhalation, this is usually followed by some slight contraction, and must not be mistaken for the dilatation which occurs at later stages in ordinary cases; at the same time it is not altogether a pleasing sign for the administrator, indicating, as it does, a tendency to syncope, and we must therefore be particularly careful in administering gas to patients of this type.   The dilatation of the pupil, too, may not appear until the stertor, convulsive twitching, intermittent breathing, &c., have become very decided.   I have, by the way, seen the presence of a glass eye considerably disconcert the administrator, and we should always accustom ourselves to test the conjunctiva and state of the pupil in *both* eyes, before arriving at any conclusion, not only on this account, but also because they may vary considerably as to the degree of sensibility, &c., still apparent.

Stertor.—It is, under ordinary circumstances, usual to describe three conditions to which this term has been applied.

First, labial or buccal stertor, or noise made by the lips in expiration—*e.g.*, in elderly people or those subject to facial paralysis.

Secondly, palatine stertor, or snoring due to vibrations of the soft palate.

Thirdly, true or laryngeal stertor, caused by vibrations of the arytænoid folds and loose structures of the larynx.

Of these it is obvious that the first cannot very well be developed in a patient whose mouth is held widely open with a mouth-prop. Snoring is a very constant phenomenon in the course of nitrous oxide inhalations, and although it is not usually considered of sufficient importance to call for remark I think some little attention should be paid to its early development and intensity: when the snoring is very loud and the palate presumably very lax, there is a greater tendency for the fauces, back of the tongue, &c., to become irritated, and consequently to give rise to retching and vomiting.

True stertor is generally developed sooner or later in the course of the inhalation, and has a peculiar irregular sound, short, spasmodic expiratory efforts follow one another in rapid succession, and are associated with a considerable amount of mucous rattling.

Spasm.—The convulsive twitching and oscillations of the eyeball are frequently very decided, but need cause no alarm. As the convulsions become more general, a condition of *clonic spasm* is gradually induced, and may cause some inconvenience to the operator, and hence the gas should be withdrawn before the spasmodic symptoms become marked. In hysterical females especially, a *tonic* instead of a clonic condition of spasm, in which the whole muscular system is involved, may be induced without previous twitching; the hands are then clenched very firmly, the legs and arms stiffened and moved across one another, the whole body being similarly rigid and in a state of more or less marked opisthotonos; pleurosthotonos or emprosthotonos may also be produced, though more rarely; if in these cases the spasm does not pass off after one or two more inspirations, but tends rather to become intensified, it is best to discontinue the administration and commence the operation, the patient being gently restrained from slipping out of the chair the while. The possible development of this spasmodic condition should also warn us against placing the mouth-props over doubtful teeth, which are liable to be broken, or even forced out of the socket, or over bare gums, and against allowing the hands to clasp either one another or any adjacent hard object.

Under this head, too, allusion must be made to the occasional development of a condition known, in this connection, as " subluxation of the inferior maxilla," by which is meant, that sometimes after taking the gas, when complete consciousness has returned, the patient is unable to close the widely open mouth. I am inclined to doubt whether any true dislocation, or even subluxation of the condyle, occurs in these cases, but look upon it as merely an exaggeration of the usual tonic spasm. I have, however, seen it so marked, as to require the manipulations usual for the reduction of actual dislocation—*i.e.*, the insertion of the thumbs into the mouth, and backward and downward pressure at the angles of the jaw. As a rule, however, simply lifting and slightly drawing forward the lower maxilla will serve the purpose. The condition in question will, I believe, be found to be usually associated with the use of the mouth-props between the incisors, especially if the jaw is at all " underhung."

Vomiting and Retching.—It is very doubtful whether actual vomiting ever occurs as a primary phenomenon during the inhalation of the gas. Retching may generally be ascribed to impurities in the gas itself, or more frequently perhaps to malposition of the head, or in second administrations to regurgitation of blood as the result of the previous operation; occasionally, the mouth-prop may slip on to the base of the tongue, and mere efforts at swallowing made by nervous patients, are often sufficient in themselves to induce retching, but it hardly seems possible that actual distension of the stomach with gas can then ensue. This retching generally disappears, unless due to a mechanical cause, upon removal of the face-piece; should it not do so, or if the mouth-prop is found to have moved, the latter should be withdrawn or held in the anterior portion of the mouth, (it is not always possible to withdraw it at once, on account of the spasmodic closure of the jaws), and the head should be lifted into the erect position, in order to throw the uvula well forward; if, in spite of this, vomiting takes place, care should be taken that none of the vomited matter regurgitates into the larynx, and, if sufficiently conscious, the head and shoulders of the patient should be pushed well forward over a bowl held in the lap, or when this is not possible, the head may be turned, so as to allow the vomited matter to run out of the mouth on one side.

Borborygmi are frequently heard, especially in females, and are sometimes associated with the passage of flatus per anum (see page 22).

Involuntary Micturition, or even Defæction, occasionally occur, either quite at the end of inhalation, or, especially in young and hysterical girls, during the excitement of recovery.

Sexual Excitement.—It has been repeatedly mentioned above, that sexual disturbance is frequently associated with inhalation of the gas, and we must be prepared for some manifestations of this condition either in the early or late stage of excitement.

Recovery.—On first removing the face-piece, it is by no means uncommon to find that the lividity, stertor, and spasm tend momentarily to increase in intensity, and the pupil, if it has not already, now almost invariably dilates. This apparent increase in the intensity of the symptoms is, in all probability, due to the continued absorption of the gas from the ultimate air vesicles of the lungs, to which the first one or two inspirations of air do not at once penetrate. The patient then goes through precisely the same stages, though of course inversely, to those described above—*i.e.*, with the renewal of the aërial respirations the narcotic gradually gives place to a stage of excitement.

Although recovery is usually very thorough and complete, it may be associated with a slight feeling of faintness, this seldom goes on to actual syncope, and is usually quite abolished, if the patient is directed to lie down on a couch with the head low for a few minutes, is kept warm, and well supplied with fresh air.

I can, too, recall cases, in which vomiting and vertigo were induced by each effort the patient made to rise from the operating chair, even ten minutes, or a quarter of an hour after complete return to consciousness. Such cases usually yield to rest in the recumbent posture, associated with a plentiful supply of fresh air.

Serious syncope and asphyxia will be discussed under the head of Dangerous Symptoms in a future chapter.

It will be gathered from what has gone before, that not only are the symptoms which follow nitrous oxide inhalation, liable to great variation in intensity, but that the exact order of their development is very uncertain. I would, therefore, insist most strongly, upon the importance of keeping a record of all cases in which anæsthetics are administered, whether the course of the administration is marked by the development of abnormal symptoms or not. Accumulated evidence, such as records of this description would alone afford, is always of value sooner or later, and would do much to clear up many uncertainties. In order to assist those who may be inclined to act upon this suggestion, I have endeavoured to lighten the labour of note-taking, by drawing up a record or register, on the plan of the well-known Obstetric Register published by Messrs. Smith, of 52, Long Acre, from whom this Anæsthetic Register can be obtained. In preparing this form I have been much assisted by the kindly advice and encouragement of my colleagues of the National Dental Hospital, to whom I would take this opportunity of returning my best thanks. To those who, having obtained one of the forms in question, may be somewhat appalled by the number of columns and separate headings, I would suggest that it is hardly to be expected that *every* column should be filled up, in each individual case, although, of course, the more complete the notes, the greater will be the value of the records.

CHAPTER VIII.

After Effects and Special Cases.

THE most usual complaints made by patients, as to the after effects of the gas, are slight headache and vertigo, and a general feeling of lassitude and depression; but these are by no means general, and if severe, should I think, lead us to suspect the quality of the gas, or defective administration, rather than any peculiarity on the part of the patient.

It is not infrequently remarked by feeble, anæmic patients, that their condition for the remainder of the day of inhalation, was distinctly better than usual, but that, on the following day the lassitude and depression was very decided; in two or three old-standing hemiplegics to whom I have administered the gas, the affected limb has appeared obviously warmer after, than before the inhalation, and this condition has been maintained for some hours. A condition of torpor, allied very closely to coma, and of considerable duration, has occurred in one or two instances. Other observers speak of the supervention of hemiplegia and temporary catalepsy.

At best, however, our knowledge of the after effects of the gas is but scanty, and in no portion of the subject would systematic records of cases be likely to prove of greater value: in obtaining such records, the hospital anæsthetist is obviously placed at a great disadvantage, compared to the general practitioner, to whom therefore, we must look for improvement in our knowledge of these obscure conditions.

SPECIAL CASES.

Although reference has been made, more especially in the chapter immediately preceding, to the variations in the phenomena, which are associated with some of the more obvious morbid processes, it will perhaps be convenient if attention be drawn more particularly, to certain special cases that may call for exceptional skill and care.

Consecutive Administrations.—It is sometimes considered advisable, for reasons into which we need not now enter, to continue, or complete immediately, an operation already commenced, but which the brief duration of the anæsthesia has not permitted the operator to finish, in the course of a single administration of the gas. The question will then arise, as to the advisability of repeating the inhalation, and the anæsthetist will be appealed to for his opinion, as to the fitness of the patient to undergo a subsequent administration; in coming to a conclusion upon the subject, the following points must be taken into consideration:—

    (a) The tendency to the development, or intensification, of variable symptoms, idiosyncrasies, and the effects of hysteria are, of course, much more marked in a second or subsequent administration.

    (b) As might be expected, unpleasant after effects are much more likely, though by no means certain to occur, especially if the first administration was followed by hysteria or syncope.

    (c) Although, by the use of anæsthetics, the shock attendant upon surgical operations is reduced to a minimum, the mere extraction of a large number of teeth, or, in fact, the performance of any operation, always gives rise to a certain amount of "after shock," which it may be advisable to avoid unless the patient can be kept under observation, *e.g.*, in bed.

If, therefore, the patient has been much "upset," to use a common but expressive term, by the first administration, we must hesitate before advising a second; but on the other hand, in an ordinary case, there need be no compunction upon the subject, providing that the patient is duly warned of the liability to some slight subsequent uneasiness.

If it is decided to re-administer the gas, the patient is allowed to become quite conscious, the mouth is well washed out, and the bleeding stopped. Beyond calling for a little extra care and vigilance on the part of the administrator, these cases do not give rise to much trouble; as a rule they become anæsthetised much sooner, and with less gas, than primary administrations, and their return to consciousness is correspondingly rapid. The gas has been given as often as six times in succession, without any apparent ill effect, but it is seldom that any necessity for such repetition occurs, and it is certainly not advisable. Some authorities even go so far as to say, that more than one administration on the same day is objectionable.

The difference between consecutive administration, as described above, and prolonged administration, as described in the chapter on the use of nitrous oxide in general surgery, must be borne in mind. In the one case, the patients are placed profoundly under the influence of the gas, twice or three times in succession, with intervals of complete recovery; in the other, the patients are not allowed to regain consciousness until they can do so permanently, after the operation has been performed.

Children.—The administration of gas to children calls for some few remarks. The admission of the friends or parents of the child to the operating room is the first difficulty; some children are more amenable in the presence of their parents, others, and I believe the majority, are not, so I prefer when possible, to dispense with them, placing the child at once in the operating chair. A brief effort should be made to introduce the mouth-prop, and a little gentle persuasion will generally effect our purpose, but if we do not succeed, do not weary the child with arguments, but proceed at once to the administration, using rather a small face-piece, and gently restraining the struggles of the little patient. When the struggles have ceased, and the child has come well under the influence of the gas, remove the face-piece, and insert the gag or mouth-opener between the teeth; if this can be done rapidly, one or more teeth can be extracted at once, but if time is lost in opening the jaws in consequence of the spasm, a mouth-prop may be slipped between the teeth, and a little more gas administered.

As a rule children pass under the influence of the gas very rapidly and deeply, and recover equally quickly; crying children especially, appear to be affected to the full physiological extent almost suddenly, owing to the deep sighing inspirations which are then taken, so that the administrator must be vigilant and careful.

Elderly People.—If in good health, people over fifty years of age take gas well, and it has frequently been given to much older patients; such people pass very quickly under the influence of the gas, the lividity is more marked, and the anæsthesia more profound, and relatively, of slightly longer duration; inasmuch too, as they are frequently subjects of senile changes, in the shape of thickened arteries, feeble hearts, and diminution of respiratory power, these effects (lividity, &c.) are then even more pronounced.

Heart Disease.—The administration of gas in known cases of cardiac disease, though not absolutely contra-indicated, should be approached with caution and carried out with care, and always by, or in the presence of, a qualified medical man. In addition to careful auscultation of the heart, the condition of the pulse in the wrist, or in some such easily available artery as the superficial temporal, should be ascertained before commencing the inhalation, in order that any changes in its rate, rhythm, and quality may be duly appreciated as the administration proceeds; for this purpose the finger should be kept applied to one or other of these spots during the whole period. The lividity generally becomes very marked at an early period of the inhalation, and the pulse slow; at the first sign of intermittence of the latter, the gas should be at once withdrawn. The anæsthetist should also take care to have at hand, tongue forceps, and capsules of nitrite of amyl especially, and it is often as well before commencing the inhalation to administer a little weak stimulant. Mere functional palpitation is not likely to cause any trouble—in fact, as might be expected, the heart's action in these cases is steadied and strengthened rather than the reverse.

Pulmonary Disease.—In tubercular conditions, associated with large cavities and considerable secretion, the loss of breathing space forms a serious impediment to the action of the gas, which should therefore, be administered with caution, as such patients rapidly show signs of asphyxia and cardiac failure: in such patients, too, the symptoms are generally intensified on removal of the face-piece (page 79). Cases of hæmoptysis subsequent to the inhalation sometimes occur. The emphysematous and bronchitic are also bad subjects for the gas; and though its use in these cases need not be forbidden, we must recollect that the right side of the heart is already overloaded, the blood deficiently aërated, and consequently, the lividity becomes intense and the heart tends to fail during inhalation.

Nervous Disorders.—Dr. Savage, the Medical Superintendent of Bethlem, in a recent paper,* has raised the question of the advisability of administering anæsthetics to those who are known to be subject to mental aberrations, and the influence of artificial anæsthesia upon insanity in general. Although in that paper, chloroform and ether are more particularly alluded to, the remarks and adverse opinion expressed must apply, though in lesser degree, to nitrous oxide.

Both epileptic and choreiform seizures are recorded as having been induced by, and as having followed inhalation, but such cases are exceptional, and do not occur sufficiently frequently, to justify us in refusing to administer the gas to the subjects of these conditions.

Pregnancy.—There does not appear to be any objection to the administration of nitrous oxide at any stage of pregnancy (except perhaps at term, when the spasm might possibly induce labour); the duration of the anæsthesia is so short that neither mother nor child appear to suffer. There is no record of the effect, if any, upon placental or fœtal growth in the earlier stages. It would almost appear probable too, that lactation is much more likely to be interfered with, by the performance of an operation *without* an anæsthetic, than it is when the gas is inhaled.

Alcoholics.—People whose tissues and organs are in a bad condition, consequent upon alcoholic excess in the past, are bad subjects for the administration of most anæsthetics: as far as nitrous oxide inhalation is concerned, no particular danger is likely to arise from this class of patients, providing, of course, that their cardiac and respiratory functions are fairly good; but in the chronic, and more especially in the

* "British Medical Journal," December 3rd, 1887.

acute forms of alcoholism, one must be on one's guard against
the exaggerated stage of excitement, both before and after
actual anæsthesia.   I have myself seen violent pugilistic
tendencies develop, in a man who had been indulging rather
freely immediately before inhalation.

Hysteria.—It would be quite useless to attempt to enu-
merate, even a tithe of the curious and untoward symptoms,
to which the condition known as hysteria is likely to give
rise, and I must content myself with the warning, that the
mental condition of a patient recovering from nitrous oxide
anæsthesia, is precisely the one in which hysterical manifestations are likely to occur.

It is equally impossible to lay down rules for the
treatment of such cases.   Providing that they do not injure
others, it is seldom that they hurt themselves, and they should
not therefore be restrained ; any exhibition of anxiety on the
part of any of the attendants, whether professional or lay, is
always inadvisable ; an ample supply of fresh, and by preference cold air, assists materially in the recovery.

CHAPTER IX.

Syncope and Asphyxia.

I HAVE preferred to retain for a separate chapter, an account
of such complications as may be termed dangerous, in con-
tradistinction to the minor difficulties referred to in the
preceding pages. The comparative immunity from fatal
consequences enjoyed by nitrous oxide, very justly entitles it
to the honour of being considered, the "safe" anæsthetic;
recorded cases of death, occurring while under the influence
of the gas, (whether due, strictly speaking, to that agent or
not), can be reckoned upon the fingers, while the number of
patients who have inhaled it, for various purposes, may be
counted by hundreds of thousands, or even millions. Without attempting to discuss, seriatim, these few fatal cases, it
may be as well to point out some of the lessons which they
teach us.

1. It would appear that when death has occurred
while under the influence of the gas, it has been
due either to syncope or asphyxia.
2. Simple faintness may occur, either before the
inhalation is complete, apparently from fright, or
during the stage of recovery, apparently from the
shock of the operation.
3. The fatal syncope has with very few exceptions,
(Dr. Clarke's case, "British Journal of Dental
Science," 1883), been distinctly traceable to shock,
consequent upon commencing, or continuing the

operation when the individuals were incompletely anæsthetised, and have, as may readily be imagined, occurred in patients more or less out of health, and therefore more susceptible; in one case in particular,* well marked cardiac lesions were found at the post-mortem. Hence the importance of "*Never permitting an operation in a semi-anæsthetised condition.*"

4. The occurrence of asphyxia from simple laryngeal spasm, or from falling together of the epiglottidean folds, has not been recorded in reference to nitrous oxide, and does not seem likely to give rise to any trouble. Such a condition would, of course, be most probable during the height of anæsthesia.

5. Fatal asphyxia has invariably been the result of some mechanical cause—*e.g.*, regurgitation of vomited matter, slipping of gag or tooth into the larynx, &c., and has therefore usually occurred during, or immediately after the completion of the operation. Hence we must carefully watch the mouth during recovery, lest stray teeth or other foreign bodies slip into the larynx.

It is important that the signs of syncope or asphyxia should, if possible, be recognised early, the question therefore arises, Is such early recognition possible? and if so, how?

SYNCOPE.

The occurrence of syncope, to whatever cause it may be referable, is usually very sudden, and cannot therefore, be said to have any "premonitory" stage; unless, perhaps, the sudden dilatation of the pupil, which occurs sometimes immediately

* "British Medical Journal," 1877, Vol. I, p. 439.

before, and is sometimes coincident with the cardiac failure, can be so termed. But we may gain a good deal of information as to the possibility of its development, by a knowledge of the constitution, or even from the mere aspect and physical appearance of our patient, before inhalation commences, and being to a certain extent forewarned by such knowledge, may do much to obviate the development of this unpleasant phenomenon. For instance, we must be prepared for cardiac failure in the anæmic, and those recovering from acute disorders, or who are debilitated by any chronic disease, *e.g.*, phthisis; in those suffering from definite cardiac lesions, and in those of a highly-strung nervous temperament. But all these, of course, are but uncertain and merely conjectural signs, and in no way to be implicitly relied upon; we must depend mainly upon our own ability to recognise the condition when it *does* arise, rather than upon possibilities of its occurrence, and I must again insist upon the fact, that such syncope is due to fright or shock, rather than to any specific effects of the gas itself.

The appearances presented by a person who has fainted, are so well-known as hardly to need description; the feeble fluttering pulse, extreme pallor, muscular relaxation, dilatation of the pupils, cold, clammy sweat, and almost complete cessation of breathing, vary but in degree in cases of simple fainting, and those of fatal heart-failure, and the former may pass very readily into the latter. In syncope arising during the course of nitrous oxide administration, the signs and symptoms are but slightly altered; the change in the colour of the face is not so marked, or partakes more of the ashy hue of death, in consequence of the pre-existing lividity. The alteration in the character of the respirations is very decided; from being hurried, noisy, or even stertorous, they suddenly appear to cease altogether, or become very shallow;

but we must bear in mind that such cessation or shallowness
of breathing, is by no means uncommon in the course of an
ordinary administration, and providing it does not continue
for more than five or six seconds, and that the pulse is not
failing, need cause no particular alarm.

Treatment of Syncope.—The cessation of respiration for
periods longer than five or six seconds, as indicated above,
usually yield very readily to simple pressure upon the
chest-wall; if by this means the breathing is not restored,
and symptoms of fainting supervene, prompt measures are
necessary in order that the condition should not pass into
one of fatal syncope.

Place the patient in such a position that the head is lower
than the body and feet, hanging over the edge of a couch or
bed is perhaps best, but flat on the floor will serve the purpose: open the mouth (if not already opened), seize the
tongue in the forceps (Fig. 21), draw it well forward, and
press forcibly upon the chest. Open the doors and windows
of the room, so that a free supply of fresh air is obtainable.
Hold a broken capsule of nitrite of amyl, or a little strong
ammonia on the stopper of a bottle, close to, but not
touching the nose and lips, so that the vapour is drawn in
on releasing the chest. At the same time, the clothing over
the chest should be rapidly removed, and the chest and face
slapped briskly with a towel dipped in cold water. In by
far the majority of cases in which syncope is developed, the
above-mentioned treatment will suffice to restore both cardiac
and respiratory action; all that will then be necessary is, to
keep the patient in the recumbent position, warm, and freely
supplied with fresh air.

If, however, these preliminary efforts meet with no
response we must act upon the supposition that the heart
continues to beat for some time after the breathing ceases,

(page 12), and must proceed without delay to perform artificial respiration; the operator, while you are at work, unloosening or cutting the clothes, so as to allow the chest free play. Three methods are generally described, those of Marshall Hall, Sylvester, and Howard : of these the first is more strictly applicable to the resuscitation of the apparently drowned, and we will, therefore, only explain the two last.

Sylvester's method.—If the patient is on the floor, slip a pillow or some form of support, *e.g.*, folded coat, beneath the shoulders, so that the head hangs down and the neck is extended. Keep the tongue well drawn forward the whole time ; an indiarubber band, over the tongue and under the chin, or some similar contrivance, will effect this admirably, and will permit the operator or assistant to employ both hands in cutting off the clothes, &c., for which, however, the anæsthetist should not wait.

Stand behind the patient, grasp the arms about midway between the shoulders and elbows, press them firmly into the sides of the thorax, rotating outwards at the same time ; maintain this position for a couple of seconds, the assistant forcing the diaphragm upwards by pressure upon the abdomen the while ; then steadily draw the arms upwards and outwards, until they nearly meet above the head, slightly lifting the patient from the ground, and, at the same time, the assistant releases the diaphragm; then repeat the downward movement, and so on, each movement upwards and downwards being repeated 15 or 16 times in a minute.

Howard's method.—With the patient on the floor, and the tongue out, kneel across the abdomen; direct the operator or assistant, to draw and hold the extended arms above the head ; grasp the margins of the thorax with the palms of the hands, the fingers towards the axillæ, the thumbs towards the xiphoid cartilage ; press upwards and inwards towards the diaphragm, gradually bending over the patient, and allowing the whole weight of your body to assist the movement; maintain this pressure for a few seconds, then briskly push yourself to the kneeling position again, and recommence the action, at about the same rate as in Sylvester's method, *i.e.*, 15 or 16 times a minute.

While the anæsthetist is working at artificial respiration, the operator or an assistant may try the effect of dashing cold water on the chest and face, and vigorously rubbing them with a dry, rough towel, or, if an electric battery is at hand, he may place one pole over the sterno-mastoid, the other over the heart. Acupuncture of the heart, by passing a fine needle into the ventricle, has been suggested, its action being probably mechanical. But valuable time should not be frittered away in such experiments; our chief reliance should be placed in artificial respiration, which should be continued perseveringly by the anæsthetist for at least an hour.

If signs, however faint, of a revival of cardiac or respiratory action are observed, our efforts should be redoubled, and in addition the lips, gums, tongue, and inner sides of the cheeks should be gently rubbed with a cloth dipped in brandy; at the same time hot bottles, mustard plasters, &c., should be applied to the soles of the feet, calves, epigastrium and præcordium, the whole body must be kept warm, and a mixture of hot beef-tea and brandy, (an ounce of each), at once injected into the rectum. We must not relax our efforts until ordinary respiratory movements are well established, and the pulse has become quite regular, and even after this we must remain with our patient for some little time.

ASPHYXIA.

Asphyxia occurring in the course of nitrous oxide administration in dental work, is, almost invariably, the result of mechanical obstruction, and as such occurs usually during the latter period of recovery: this is an important point, for it is on the *recurrence*, or intensification, of the lividity, that we depend for the early recognition of the impending crisis.

We cannot, as in syncope, be in any way forewarned by the appearance of our patient prior to administration; it is, therefore, most important that the anæsthetist, during the stage of recovery, should not relax his vigilance, but keep his eyes upon the face of the patient; and not only observe the lips, but also notice the direction the blood is taking; that the gag does not slip, and that the teeth are safely removed from the mouth as extracted. It must also be remembered that with the mouth widely open, the tongue tends, if the patient is recumbent or the head tilted, to fall back over the laryngeal opening.

The signs of commencing suffocation are : the increased or returning lividity, which extends very rapidly over the whole surface, even to the fingers ; the gasping and struggling for breath, terminating in actual convulsions, and cessation of respiration. The violent inspiratory efforts, as well as the non-oxygenation of the blood, themselves act as cardiac depressants, and the heart's action is seriously impeded, and, finally, stops.

It is most important to recollect that the actual chest movements may continue in spite of the fact that the glottis is completely occluded, and we must not, therefore, depend upon these alone for evidence that air is entering the lungs.

Treatment of Asphyxia.—If we have reason to suspect that a stray tooth, a mouth-prop, or other hard foreign body, is in danger of slipping into the larynx, it is undoubtedly the duty of the anæsthetist, by sweeping his finger round the mouth, to attempt to avert the accident ; a special instrument termed Carter's spoon, made of wire gauze in the shape of a bowl of a spoon, has been introduced for the purpose of guarding the larynx during the extraction of teeth, but it may

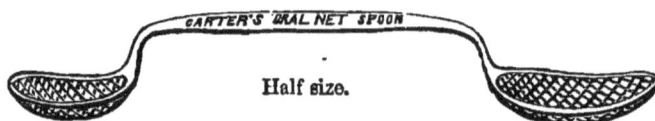

CARTER'S ORAL NET SPOON

Half size.

FIG. 26.

seriously impede the movements of the operator, and so has not come into very general use.

If signs of asphyxia from this cause have already developed, it is not at all a bad plan, especially in children, to seize the patient by the heels, lift them up, and allow them to hang head downwards over the back of the chair, and so encourage the foreign body to fall out of the mouth by its own weight. This method of inversion is hardly possible with adults, and may lead to some loss of valuable time. If the foreign body is in sight, or within reach, endeavour to remove it with the finger, or by the help of the throat forceps (Fig. 22), keeping touch of it the whole time. In by far the majority of cases, the offending substance will be quickly brought up by the violent cough to which irritation of the glottis gives rise; should this favourable termination not ensue, and the signs of asphyxia become on the contrary more marked, we must have recourse to laryngotomy. The patient must be laid upon the floor with the shoulders raised, and the neck extended, and the trachea must be opened between the thyroid and cricoid cartilages, according to the directions given in surgical works; this being done, artificial respiration, according to one or other of the methods already described must be employed until the breathing is restored; intubation of the larynx is not sufficiently well understood in this country, to warrant one in suggesting its trial in these cases, but it would, no doubt, be of considerable service if skilfully and rapidly performed.

In cases such as described above, where the asphyxia is undoubtedly due to the presence of a foreign body, whether obstructing the entrance of air by its size, or causing spasm of the glottis by irritation, do not be in a hurry to draw the tongue forward, and so open the way to the larnyx.

If we are suspicious that asphyxia due to the presence of blood and mucus in the larnyx is supervening, and have no reason to think that it arises from the presence of a foreign body, e.g., a tooth, our first manœuvre is to seize the tongue and draw it well through the teeth, at the same time, if the patient is sitting, lift the head and shoulders, and push them forward, so that the blood may flow out of the mouth; it is then a good plan to clear away any clotted blood and mucus that may be aggravating, if not actually causing the mischief, by thrusting a good-sized, dry, coarse sponge (which may or may not be attached to a stick), as far into the pharynx as possible, withdrawing it rapidly with a sweeping movement, but this must not be attempted if the obstruction is due, or suspected to be due, to a foreign body which would then be thrust more firmly on to, or even through the glottis. If, in spite of our efforts, respirations cease and lividity increases, we must be prepared to treat the case exactly on the same lines as suggested on the previous page. If we are doubtful whether the obstruction is due to blood and mucus, or to some hard substance, we ought, I think, first to act upon the former supposition, and endeavour by pulling forward the tongue, clearing away the mucus, &c., to elimi- nate that cause before proceeding further; if no relief follows, we may pass the index finger well into the back of the throat, and ascertain if any hard substance is within reach, but we should not, I am sure, be justified in groping about the

H

neighbourhood of the glottis with the throat forceps, at the risk of severe and permanent injury to the delicate structures connected with the larynx, unless we could either see or feel the obstructing body, and of course we should have no time for laryngoscopic examinations.

Although mechanical obstruction from such accidents as those above mentioned—*i.e.*, blood and mucus, and hard foreign bodies—has been the sole cause of asphyxia in the hitherto recorded fatal cases, it is only right to point out, that partial asphyxia *may* arise in nitrous oxide anæsthesia, from the folding together of the arytæno-epiglottidean folds, as was pointed out by Sir Joseph Lister in reference to chloroform; if this occurs, dangerous symptoms similar to those described above, may supervene quite suddenly without even preliminary stertor. Recovery is almost equally sudden if the tongue is drawn forcibly out of the mouth ; the folds then recede, probably in great measure owing to the reflex action, induced by the irritation of the frenum linguæ against the teeth, and of the forceps holding the tongue. Fatal effects ought not to follow in these cases, and will not do so if we recognise the position of affairs early and apply the proper remedy without delay.

Happily, kindly nature comes to our aid in by far the majority of these cases, and the cough arising from irritation of the glottis, by the foreign body itself, assisted, maybe, by a few vigorous slaps on the back, is most frequently successful in removing the source of trouble. I would, therefore, insist most strongly upon the advisability, not only of being prepared for emergencies, but also of choosing the right time for interference, of avoiding undue hurry and bustle, and of assisting rather than attempting to supplant the natural efforts.

After Effects of Foreign Bodies.—The dangers likely to arise in cases in which foreign bodies have slipped into the larynx, are not only those of immediate asphyxia, but also the more remote ones consequent upon the passage of the foreign body *through* the glottis, and its lodgment in the trachea, or one of the bronchi, leading either to death, after weeks or months of suffering, or prolonged illness terminating in the removal of the substance, either spontaneously or by surgical operation.

The importance of the subject must be my excuse for attempting, at the risk of tedious repetition, to summarise the remarks contained in the above pages.

SYNCOPE.

Recognised by—Sudden dilatation of pupil, extreme pallor, heart failure, muscular relaxation, feebleness of breathing.

Treatment.—Prone position, draw out the tongue; artificial respiration.

ASPHYXIA.

Recognised by—Increasing duskiness, violent efforts at respiration, gradual failure of pulse.

Treatment.—Remove foreign bodies (inversion), or draw the tongue *well* forward, press on chest, wipe out mucus; laryngotomy followed by artificial respiration.

CHAPTER X.

Nitrous Oxide in General Surgery.

Considering the intimate connection, both historically and
clinically, between nitrous oxide and dentistry, it will not be
surprising, that the illustrations and descriptions of its use
and action contained in the previous pages, should have been
taken more particularly from that department of the profession.
But its application is by no means confined to dental surgery.

For short operations and minor surgery in general, the
use of the gas comes within the range of the general practitioner, who is also liable to be called upon to administer
it for members of the dental profession, and it is therefore
highly desirable that he should make himself acquainted with
its action, method of administration, and capabilities in
general; and it would, I am sure, be used even more frequently,
were a knowledge of its properties and of the apparatus
employed more universal. Unfortunately, the whole subject
of anæsthetics is much neglected in our schools, in spite of
the fact that it is in great measure to their discovery that
many of the most notable improvements, in surgery especially,
are due. It is only at comparatively few schools, (I refer more
particularly to medical schools), that even the slightest attempt
is made to give systematic instruction upon the subject; in
many, I had almost said the majority, a student may pass
through the whole of his course, and receive his diploma, not
only without seeing nitrous oxide itself administered, but

without ever having one single word addressed to him upon anæsthetics of any kind. Taking into consideration the fact, that no inconsiderable number of deaths are recorded from year to year, as being due to the administration of anæsthetics in some form or another, it is to be hoped that the authorities in whose hands rests the drawing up of the curriculum, will, at no very distant date, require from candidates evidence of having obtained, either by actual experience or from instruction received, a certain amount of information upon the subject.

The primary objects we have in view in administering an anæsthetic to a patient are—

      1. Alleviation of pain ;
      2. Abolition of shock ;

and as far as these are concerned, nitrous oxide may be said to fulfil all requirements. In certain operations, however, notably those connected with the abdominal cavity, absolute quietude on the part of the patient, with relaxation of all muscular spasm, is almost equally essential, and for these the gas is obviously not so suitable. Other objections which have been urged against its use in general surgery are—

1. The shortness of the anæsthetic stage ; but though the period of anæsthesia is certainly very brief, it can, by a little judicious management, be prolonged for a time at least sufficiently long, to perform many of the simpler operations of minor surgery.

2. That rather more special skill is required in its use, and that the apparatus is, at present, somewhat heavy and costly, both of which drawbacks are, I venture to think, more apparent than real ; the latter especially will, I trust, not be allowed to have too much weight where the safety and comfort of our fellow-creatures is concerned.

These, as far as I know, are the only objections which have been seriously raised to its use; but, on the other hand, the advantages of nitrous oxide over other forms of anæsthesia are so great, that the *possibility* of its employment, in some form, should always be considered, when the question of performing a short operation under an anæsthetic arises; these advantages I take to be—

1. The accidents which have arisen during its use are so few as to entitle it to be termed, the " safe " anæsthetic.
2. No previous preparation of the patient (such as starving) is required.
3. It can be administered with equal safety in any position of the patient.
4. It is pleasant to take, and quick in its operation.
5. Recovery is rapid in the extreme, and seldom, if ever, followed by disagreeable after effects, and almost never by the troublesome nausea and sickness of chloroform.

In Minor Surgery.

Many small operations in which the gas has been or is likely to be useful, will occur at once to my medical readers : as a diagnostic agent, too, I am inclined to think that nitrous oxide might be used much more freely than it is.

I will now attempt to point out *some* of its most obvious applications, and describe the methods of administration, in so far as they differ from those alluded to in the previous chapters.

Opening *abscesses* in the subcutaneous and cellular tissues, especially if they are associated with much inflam-

matory thickening and tension. Under this head of course
would come such special abscesses as gumboils, carbuncles,
whitlows, ischio-rectal abscess, &c.

Applications of *actual cautery* to chronically enlarged
joints, spine, &c.

Breaking down *adhesions in joints* that have become stiff.

Boils and abscesses in the external *auditory meatus* are
usually very tender and painful and require free incision,
and the gas is then very useful, as it is essential in these
cases that the patient should not start or shrink from the
surgeon's knife.

In painful, or forcible *catheterisation*, and sounding for
stone, when the urethra is particularly sensitive.

Removal, or incision and scraping, of so-called sebaceous
cysts, and small dermoid tumours, which would include bursæ
such as "housemaids' knee," compound ganglia, hæmatomæ,
enlarged glands, &c.

In the use of the *Eustachian catheter* when the nasal
passages are hyper-sensitive, ulcerated, or inflamed.

Cutting down upon and removing such simple *foreign
bodies* as splinters of wood and glass, needles, &c., which may
have become embedded in the tissues.

Removal of small *hæmorrhoids* by clamp and cautery.

Avulsion of *in-growing toe-nail*.

Internal urethrotomy, especially when the stricture is
situated within a short distance of the meatus.

In the examination of painful *joints*, especially in hysterical or neurotic patients.

Incision of the *membrana tympani*, which requires very
careful and delicate manipulation, is much facilitated by the
administration of the gas.

I have not alluded to *ophthalmic operations*, because this
is the special department of cocaine, but it is obvious that

there are many operations which could be done under the gas, *e.g.*, the removal of foreign bodies embedded in the cornea, though the attendant oscillations of the eyeball are sometimes objectionable. It may also be used in operations upon the canaliculi, *e.g.*, passage of probes and actual division.

Removal of foreign bodies and *polypi* from the nose and ear, in which cases cocaine is by no means certain in its action.

In the examination of the throat, and of naso-pharynx for *post-nasal growths*, &c., especially in children.

Opening *prostatic abscesses* from the rectum.

In *rectal examinations*, when the existence of fissures, ulcers, &c., causes considerable pain.

Inserting *setons*.

Divisions and dilatations of simple *sinuses*.

Tenotomy, and division of cicatrices and contracted fasciæ, though the spasm induced is sometimes held to be objectionable.

Tonsillotomy, and removal of uvula.

Scraping unhealthy *ulcers*, *lupus*, &c., with a sharp spoon, frequently a very painful process.

The use of nitrous oxide by itself is obviously not possible, or even advisable, in certain surgical proceedings, such, for instance, as in the reduction of dislocations or herniæ, or the setting of complicated fractures, or in fact any manipulations in which the absolute relaxation of the muscles is necessary ; nor perhaps, is it wise, to administer the gas by itself to hysterical females with a view to examinations of, or operations upon the external genitals, the erotism is often very marked in such cases, and may lead to somewhat troublesome hysterical after effects. Its use, too, in operations for fistula and fissure of the rectum, where profound anæsthesia and muscular relaxation is required, is hardly advisable.

A word now as to the methods of administration in such cases as mentioned above. The principles involved are, of course, precisely similar, and to all intents and purposes the methods are the same as in dental work, but the following hints may be of service to those who intend to employ this agent, either in surgical operations or in diagnosis.

1. The patient should be placed and restrained in the exact position in which the operation is to be performed, or the examination made. To say nothing of the shortness of time, it is exceedingly difficult to move a patient whose muscles have been thrown into a state of spasm by nitrous oxide.

2. A mouth-prop *need* not be introduced unless, of course, the operation is to be in the buccal cavity itself, but it is often of advantage, as the oval face-piece will then fit better. If no prop is inserted, the more conical face-piece (Fig. 12), rather smaller than usually employed in dentistry, is used.

3. When the mouth is closed, patients take a little longer to pass under the influence of the gas than when it is propped open; and, especially if recumbent, the snoring and stertor may be delayed and rather less marked. In other respects the phenomena, &c., are much the same as described in Chapter VI.

4. If necessary, the prolongation of the anæsthesia may be obtained by first of all placing the patient well under the influence of the gas, then removing the face-piece for one, or at most two, aërial inspirations, quickly replacing for an equal number, and so on alternately removing and re-applying for as long as may be necessary or advisable, particular attention being paid to the state of the pulse the whole time. M. Bert's experiments with mixtures of nitrous oxide and oxygen at ordinary pressure, as a means of prolonging the anæsthesia induced by nitrous oxide alone (see page 30), were

but developments of this system. It has also been suggested to continue the inhalation, in operations about the
mouth, by passing through the nose into the pharynx a small
tube—*e.g.*, a gum elastic catheter, No. 8—connected with the
reservoir bag. The respirations then consist mainly of air
mixed with a small proportion of nitrous oxide, but this
method has not been extensively tried, and does not seem
likely to be of much service. Of course prolongations,
however attempted, are the source of much anxiety to the
anæsthetist, and call for a degree of mental and physical con-
centration and strain which, to say the least, are exceedingly
trying. They should not, therefore, be undertaken lightly
and without fully understanding and appreciating the grave
responsibility assumed. With this proviso there is no particular reason why they should not be attempted.

In Major Operations and those requiring some time for their Performance.

The days of "brilliant" surgery and "lightning" operations have passed, and it is seldom considered any particular
merit for a surgeon to amputate a limb, remove a tumour,
&c., in any surprisingly short space of time; on the
contrary, painstaking and scientific surgery, which is .often
synonymous with prolonged operations, is the order of the
day, and such surgery obviously does not lend itself to the
use of nitrous oxide alone as the anæsthetic agent to be
employed. In spite of this, however, many cases are on
record where, either designedly or from accidental complications occurring in the course of an originally simple operation,
it has been found necessary to maintain its influence for
very prolonged periods. One of the earliest operations
performed with the gas was an excision of the breast, (by

Dr. Bigelow, in America, in 1848), and it has even been administered continuously for upwards of 30 minutes. It is, of course, seldom, if ever, that one would commence the inhalation of nitrous oxide with the idea of maintaining it alone for such a length of time; reference has been made to these cases rather with a view to emphasise the capabilities of the gas, than as examples which it is at all desirable to imitate.

I have already alluded, (page 29), to M. Paul Bert's experiments with nitrous oxide and oxygen under pressure, and to the special chamber necessary to carry out his views; however hopeful one may be as to the future possibilities of his methods, they can hardly be said to have arrived at a very practical stage.

The most frequent use made of the gas in lengthy operations is as an introduction, or adjunct, to the administration of ether, and with these ends in view it has proved invaluable. Without going very deeply into this question, it may be stated briefly that the advantages claimed for what is termed the "combined method" are—

1. The avoidance or modification of the preliminary excitement and spasm due to ether alone.
2. The shorter time required to induce anæsthesia, which is, however, equally profound and satisfactory.
3. Mitigation of the troublesome after-sickness.
4. Greater comfort to the patient, to whom the taste and smell of ether is usually very unpleasant.

Against these obvious advantages we have to place the comparatively minor one, that special forms of apparatus, and consequently special training in their use, are necessary.

The matter cannot be considered to come within the

scope of such a work as this, and I shall therefore content myself with indicating the methods employed, without discussing their respective merits or peculiar advantages.

The methods are three in number, viz. :—

a. In America especially, it is sought to prolong the anæsthesia of nitrous oxide, by the admixture of the gas with the vapour derived from minute quantities of ether, (chloroform, or chloroform and alcohol being occasionally substituted). For this purpose the gas, on coming from the bottle, passes through a box or chamber to which a drop bottle containing ether is attached in such a manner, that one or more drops of ether can, at the will of the adminis-trator, be allowed to fall into the chamber, where it becomes vaporised, mixes with the gas, and passes out into the reservoir bag, so that, from almost the first, the patient inhales the gas mixed with a small quantity of ether vapour.

b. Three or four respirations of pure gas are first allowed, the ether vapour is then admitted in gradually increasing quantities, so that from quite an early period the patient breathes and continues to breathe a mixture of nitrous oxide and ether vapour in inverse proportions. In this method one or other of the special forms of apparatus invented by the late Mr. Clover is necessary.

c. The patient is first placed fully under the influence of the gas; the face-piece is then very rapidly changed for the ether inhaler, and a full supply of ether admitted ; or a sponge saturated with ether is fixed in a supplemental bag (page 41), in such a way that while the tap of the bag is turned off no

vapour of ether reaches the face-piece, but on turning on the tap towards the end of the inhalation
the gas is thoroughly saturated; by these means the
patient is, so to speak, taken unawares.

INDEX.

I

114INDEX.

PAGE

Faintness after administration 80
„ during inhalation 91
False teeth as cause of asphyxia 95
„ „ removal of 52, 61
Fasciæ, division of, under gas 104
Fatal syncope 89
First operation under gas 8
Fontaine's chamber 29
Forceps, throat 45
„ tongue 45
Foreign bodies, after effects of 99
„ „ as a cause of asphyxia 94
„ „ gas in extraction of, from ear, eye, or nose 104
Forms for recording cases 80
Formulæ, chemical 3
Frankland's analysis of gas 14
French synonyms.... 1
Friends of patients, position of 54
Functional changes in nervous system 20

GAGS, forms of 45
„ used 69
Ganglia, gas in operations on 103
Gaseous interchange, in blood 19
„ „ in lungs 14
Gasometer, advantages of 33
„ described 33
„ disadvantages of 34
„ figured 56
„ used 55, 71
General physiological effects 11
„ surgery, gas in 100–109
German synonyms 2
Glass eye as source of error 76
Gloomy prognostications of Laffont 50
Glycosuria, effects in 50
Growths, post-nasal, gas in detection of 104
Gumboils, opened under gas 103

HÆMATOMÆ, gas in operations on 103
Hæmoptysis after administration 86
Hæmorrhoids, gas in removal of 103
Hall's method of artificial respiration 93
Hallucinations after and before inhalation 74, 75
Hands, position of, during inhalation 61
Head, position of, during inhalation 61
Headache after administration 82
Heart disease, administration in 86
Heart, effects of gas on 16
Hemiplegia after administration 82
Hemiplegics, how affected by gas 82
Hermann, researches of 9
Hewitt, Dr., researches of 66, 70
History 7
Horace Wells, use of gas by 8

116   INDEX.

HARRISON AND SONS, PRINTERS IN ORDINARY TO HER MAJESTY, ST. MARTIN'S LANE, LONDON.

S E L E C T I O N

FROM

J. & A. CHURCHILL'S GENERAL CATALOGUE

COMPRISING

ALL RECENT WORKS PUBLISHED BY THEM

ON THE

ART AND SCIENCE OF MEDICINE

N.B.—As far as possible, this List is arranged in the order in which medical study is usually pursued.

J. & A. CHURCHILL publish for the following Institutions and Public Bodies:—

ROYAL COLLEGE OF SURGEONS.
CATALOGUES OF THE MUSEUM.
Twenty-three separate Catalogues (List and Prices can be obtained of J. & A. CHURCHILL).

GUY'S HOSPITAL.
REPORTS BY THE MEDICAL AND SURGICAL STAFF.
Vol. XXIX., Third Series. 7s. 6d.
FORMULÆ USED IN THE HOSPITAL IN ADDITION TO THOSE IN THE B.P. 1s. 6d.

LONDON HOSPITAL.
PHARMACOPŒIA OF THE HOSPITAL. 3s.

ST. BARTHOLOMEW'S HOSPITAL.
CATALOGUE OF THE ANATOMICAL AND PATHOLOGICAL MUSEUM. Vol. I.—Pathology. 15s. Vol. II.—Teratology, Anatomy and Physiology, Botany. 7s. 6d.

ST. GEORGE'S HOSPITAL.
REPORTS BY THE MEDICAL AND SURGICAL STAFF.
The last Volume (X.) was issued in 1880. Price 7s. 6d.
CATALOGUE OF THE PATHOLOGICAL MUSEUM. 15s.
SUPPLEMENTARY CATALOGUE (1882). 5s.

ST. THOMAS'S HOSPITAL.
REPORTS BY THE MEDICAL AND SURGICAL STAFF
Annually. Vol. XVI., New Series. 7s. 6d.

MIDDLESEX HOSPITAL.
CATALOGUE OF THE PATHOLOGICAL MUSEUM. 12s.

WESTMINSTER HOSPITAL.
REPORTS BY THE MEDICAL AND SURGICAL STAFF.
Annually. Vol. III. 6s.

ROYAL LONDON OPHTHALMIC HOSPITAL.
REPORTS BY THE MEDICAL AND SURGICAL STAFF.
Occasionally. Vol. XII., Part I. 5s.

OPHTHALMOLOGICAL SOCIETY OF THE UNITED KINGDOM.
TRANSACTIONS. Vol. VII. 12s. 6d.

MEDICO-PSYCHOLOGICAL ASSOCIATION.
JOURNAL OF MENTAL SCIENCE. Quarterly. 3s. 6d.

PHARMACEUTICAL SOCIETY OF GREAT BRITAIN.
PHARMACEUTICAL JOURNAL AND TRANSACTIONS.
Every Saturday. 4d. each, or 20s. per annum, post free.

BRITISH PHARMACEUTICAL CONFERENCE.
YEAR BOOK OF PHARMACY. 10s.

BRITISH DENTAL ASSOCIATION.
JOURNAL OF THE ASSOCIATION AND MONTHLY REVIEW OF DENTAL SURGERY.
On the 15th of each Month. 6d. each, or 7s. per annum, post free.

A SELECTION

FROM

J. & A. CHURCHILL'S GENERAL CATALOGUE,

COMPRISING

ALL RECENT WORKS PUBLISHED BY THEM ON THE ART AND SCIENCE OF MEDICINE.

N.B.—*J. & A. Churchill's Descriptive List of Works on Chemistry, Materia Medica, Pharmacy, Botany, Photography, Zoology, the Microscope, and other Branches of Science, can be had on application.*

Practical Anatomy :

A Manual of Dissections. By CHRISTOPHER HEATH, Surgeon to University College Hospital. Seventh Edition. Revised by RICKMAN J. GODLEE, M.S. Lond., F.R.C.S., Teacher of Operative Surgery, late Demonstrator of Anatomy in University College, and Surgeon to the Hospital. Crown 8vo, with 24 Coloured Plates and 278 Engravings, 15s.

Wilson's Anatomist's Vade-Mecum. Tenth Edition. By GEORGE BUCHANAN, Professor of Clinical Surgery in the University of Glasgow; and HENRY E. CLARK, M.R.C.S., Lecturer on Anatomy at the Glasgow Royal Infirmary School of Medicine. Crown 8vo, with 450 Engravings (including 26 Coloured Plates), 18s.

Braune's Atlas of Topographical Anatomy, after Plane Sections of Frozen Bodies. Translated by EDWARD BELLAMY, Surgeon to and Lecturer on Anatomy, &c., at, Charing Cross Hospital. Large Imp. 8vo, with 34 Photo-lithographic Plates and 46 Woodcuts, 40s.

An Atlas of Human Anatomy. By RICKMAN J. GODLEE, M.S., F.R.C.S., Assistant Surgeon and Senior Demonstrator of Anatomy, University College Hospital. With 48 Imp. 4to Plates (112 figures), and a volume of Explanatory Text, 8vo, £4 14s. 6d.

Harvey's (Wm.) Manuscript Lectures. Prelectiones Anatomiæ Universalis. Edited, with an Autotype reproduction of the Original, by a Committee of the Royal College of Physicians of London. Crown 4to, half bound in Persian, 52s. 6d.

Anatomy of the Joints of Man.

By HENRY MORRIS, Surgeon to, and Lecturer on Anatomy and Practical Surgery at, the Middlesex Hospital. 8vo, with 44 Lithographic Plates (several being coloured) and 13 Wood Engravings, 16s.

Manual of the Dissection of the Human Body. By LUTHER HOLDEN, Consulting Surgeon to St. Bartholomew's Hospital. Edited by JOHN LANGTON, F.R.C.S., Surgeon to, and Lecturer on Anatomy at, St. Bartholomew's Hospital. Fifth Edition. 8vo, with 208 Engravings. 20s.

By the same Author.

Human Osteology.

Seventh Edition, edited by CHARLES STEWART, Conservator of the Museum R.C.S., and R.W. REID, M.D., F.R.C.S., Lecturer on Anatomy at St. Thomas's Hospital. 8vo, with 59 Lithographic Plates and 75 Engravings. 16s.

Also.

Landmarks, Medical and Surgical. Fourth Edition. 8vo, 3s. 6d.

The Student's Guide to Surgical Anatomy. By EDWARD BELLAMY, F.R.C.S. and Member of the Board of Examiners. Third Edition. Fcap. 8vo, with 81 Engravings. 7s. 6d.

Diagrams of the Nerves of the Human Body, exhibiting their Origin, Divisions, and Connections, with their Distribution to the Various Regions of the Cutaneous Surface, and to all the Muscles. By W. H. FLOWER, C.B., F.R.S., F.R.C.S. Third Edition, with 6 Plates. Royal 4to, 12s.

General Pathology:

An Introduction to. By JOHN BLAND SUTTON, F.R.C.S., Sir E. Wilson Lecturer on Pathology, R.C.S.; Assistant Surgeon to, and Lecturer on Anatomy at, Middlesex Hospital. 8vo, with 149 Engravings, 14s.

Atlas of Pathological Anatomy.

By Dr. LANCEREAUX. Translated by W. S. GREENFIELD, M.D., Professor of Pathology in the University of Edinburgh. Imp. 8vo, with 70 Coloured Plates, £5 5s.

A Manual of Pathological Anatomy.

By C. HANDFIELD JONES, M.B., F.R.S., and E. H. SIEVEKING, M.D., F.R.C.P. Edited by J. F. PAYNE, M.D., F.R.C.P., Lecturer on General Pathology at St. Thomas's Hospital. Second Edition. Crown 8vo, with 195 Engravings, 16s.

Post-mortem Examinations:

A Description and Explanation of the Method of Performing them, with especial reference to Medico-Legal Practice. By Prof. VIRCHOW. Translated by Dr. T. P. SMITH. Second Edition. Fcap. 8vo, with 4 Plates, 3s. 6d.

The Human Brain:

Histological and Coarse Methods of Research. A Manual for Students and Asylum Medical Officers. By W. BEVAN LEWIS, L.R.C.P. Lond., Medical Superintendent, West Riding Lunatic Asylum. 8vo, with Wood Engravings and Photographs, 8s.

Manual of Physiology:

For the use of Junior Students of Medicine. By GERALD F. YEO, M.D., F.R.C.S., Professor of Physiology in King's College, London. Second Edition. Crown 8vo, with 318 Engravings, 14s.

Principles of Human Physiology.

By W. B. CARPENTER, C.B., M.D., F.R.S. Ninth Edition. By HENRY POWER, M.B., F.R.C.S. 8vo, with 3 Steel Plates and 377 Wood Engravings, 31s. 6d.

Elementary Practical Biology:

Vegetable. By THOMAS W. SHORE, M.D., B.Sc. Lond., Lecturer on Comparative Anatomy at St. Bartholomew's Hospital. 8vo, 6s.

A Text-Book of Medical Physics,

for Students and Practitioners. By J. C. DRAPER, M.D., LL.D., Professor of Physics in the University of New York. With 377 Engravings. 8vo, 18s.

Medical Jurisprudence:

Its Principles and Practice. By ALFRED S. TAYLOR, M.D., F.R.C.P., F.R.S. Third Edition, by THOMAS STEVENSON, M.D., F.R.C.P., Lecturer on Medical Jurisprudence at Guy's Hospital. 2 vols. 8vo, with 188 Engravings, 31s. 6d.

By the same Authors.

A Manual of Medical Jurisprudence.

Eleventh Edition. Crown 8vo, with 56 Engravings, 14s.

Also.

Poisons,

In Relation to Medical Jurisprudence and Medicine. Third Edition. Crown 8vo, with 104 Engravings, 16s.

Lectures on Medical Jurisprudence.

By FRANCIS OGSTON, M.D., late Professor in the University of Aberdeen. Edited by FRANCIS OGSTON, Jun., M.D. 8vo, with 12 Copper Plates, 18s.

The Student's Guide to Medical Jurisprudence.

By JOHN ABERCROMBIE, M.D., F.R.C.P., Lecturer on Forensic Medicine to Charing Cross Hospital. Fcap. 8vo, 7s. 6d.

Microscopical Examination of Drinking Water and of Air.

By J. D. MACDONALD, M.D., F.R.S., Ex Professor of Naval Hygiene in the Army Medical School. Second Edition. 8vo, with 25 Plates, 7s. 6d.

Pay Hospitals and Paying Wards throughout the World.

By HENRY C. BURDETT. 8vo, 7s.

By the same Author.

Cottage Hospitals — General, Fever, and Convalescent:

Their Progress, Management, and Work. Second Edition, with many Plans and Illustrations. Crown 8vo, 14s.

Hospitals, Infirmaries, and Dispensaries:

Their Construction, Interior Arrangement, and Management; with Descriptions of existing Institutions, and 74 Illustrations. By F. OPPERT, M.D., M.R.C.P.L. Second Edition. Royal 8vo, 12s.

Hospital Construction and Management.

By F. J. MOUAT, M.D., Local Government Board Inspector, and H. SAXON SNELL, Fell. Roy. Inst. Brit. Architects. In 2 Parts, 4to, 15s. each; or, the whole work bound in half calf, with large Map, 54 Lithographic Plates, and 27 Woodcuts, 35s.

Public Health Reports.

By Sir JOHN SIMON, C.B., F.R.S. Edited by EDWARD SEATON, M.D., F.R.C.P. 2 vols. 8vo, with Portrait, 36s.

A Manual of Practical Hygiene.
By E. A. PARKES, M.D., F.R.S. Seventh Edition, by F. DE CHAUMONT, M.D., F.R.S., Professor of Military Hygiene in the Army Medical School. 8vo, with 9 Plates and 101 Engravings, 18s.

A Handbook of Hygiene and Sanitary Science.
By GEO. WILSON, M.A., M.D., F.R.S.E., Medical Officer of Health for Mid-Warwickshire. Sixth Edition. Crown 8vo, with Engravings. 10s. 6d.

By the same Author.

Healthy Life and Healthy Dwellings:
A Guide to Personal and Domestic Hygiene. Fcap. 8vo, 5s.

Sanitary Examinations
Of Water, Air, and Food. A Vade-Mecum for the Medical Officer of Health. By CORNELIUS B. FOX, M.D., F.R.C.P. Second Edition. Crown 8vo, with 110 Engravings, 12s. 6d.

Epidemic Influences:
Epidemiological Aspects of Yellow Fever and of Cholera. The Milroy Lectures. By ROBERT LAWSON, LL.D., Inspector-General of Hospitals. 8vo, with Maps, Diagrams, &c., 6s.

Detection of Colour-Blindness and Imperfect Eyesight.
By CHARLES ROBERTS, F.R.C.S. Second Edition. 8vo, with a Table of Coloured Wools, and Sheet of Test-types, 5s.

Illustrations of the Influence of the Mind upon the Body in Health and Disease:
Designed to elucidate the Action of the Imagination. By D. H. TUKE, M.D., F.R.C.P., LL.D. Second Edition. 2 vols. crown 8vo, 15s.

By the same Author.

Sleep-Walking and Hypnotism.
8vo, 5s.

A Manual of Psychological Medicine.
With an Appendix of Cases. By JOHN C. BUCKNILL, M.D., F.R.S., and D. HACK TUKE, M.D., F.R.C.P. Fourth Edition. 8vo, with 12 Plates (30 Figures) and Engravings, 25s.

Mental Affections of Childhood and Youth
(Lettsomian Lectures for 1887, &c.). By J. LANGDON DOWN, M.D., F.R.C.P., Senior Physician to the London Hospital. 8vo, 6s.

Mental Diseases:
Clinical Lectures. By T. S. CLOUSTON, M.D., F.R.C.P. Edin., Lecturer on Mental Diseases in the University of Edinburgh. Second Edition. Crown 8vo, with 8 Plates (6 Coloured), 12s. 6d.

Private Treatment of the Insane as Single Patients.
By EDWARD EAST, M.R.C.S., L.S.A. Crown 8vo, 2s. 6d.

Manual of Midwifery.
By ALFRED L. GALABIN, M.A., M.D., F.R.C.P., Obstetric Physician to, and Lecturer on Midwifery, &c. at, Guy's Hospital. Crown 8vo, with 227 Engravings, 15s.

The Student's Guide to the Practice of Midwifery.
By D. LLOYD ROBERTS, M.D., F.R.C.P., Lecturer on Clinical Midwifery and Diseases of Women at the Owens College; Obstetric Physician to the Manchester Royal Infirmary. Third Edition. Fcap. 8vo, with 2 Coloured Plates and 127 Wood Engravings, 7s. 6d.

Lectures on Obstetric Operations:
Including the Treatment of Hæmorrhage, and forming a Guide to the Management of Difficult Labour. By ROBERT BARNES, M.D., F.R.C.P., Consulting Obstetric Physician to St. George's Hospital. Fourth Edition. 8vo, with 121 Engravings, 12s. 6d.

By the same Author.

A Clinical History of Medical and Surgical Diseases of Women.
Second Edition. 8vo, with 181 Engravings, 28s.

Clinical Lectures on Diseases of Women:
Delivered in St. Bartholomew's Hospital, by J. MATTHEWS DUNCAN, M.D., LL.D., F.R.S. Third Edition. 8vo, 16s.

Notes on Diseases of Women:
Specially designed to assist the Student in preparing for Examination. By J. J. REYNOLDS, L.R.C.P., M.R.C.S. Third Edition. Fcap. 8vo, 2s. 6d.

By the same Author.

Notes on Midwifery:
Specially designed for Students preparing for Examination. Second Edition. Fcap. 8vo, with 15 Engravings, 4s.

A Manual of Obstetrics.
By A. F. A. KING, A.M., M.D., Professor of Obstetrics, &c., in the Columbian University, Washington, and the University of Vermont. Third Edition. Crown 8vo, with 102 Engravings, 8s.

Intra-Uterine Death:
(Pathology of). Being the Lumleian Lectures, 1887. By WILLIAM O. PRIESTLEY, M.D., F.R.C.P., LL.D., Consulting Physician to King's College Hospital. 8vo, with 3 Coloured Plates and 17 Engravings, 7s. 6d.

The Student's Guide to the Diseases of Women. By ALFRED L. GALABIN, M.D., F.R.C.P., Obstetric Physician to Guy's Hospital. Fourth Edition. Fcap. 8vo, with 94 Engravings, 7s. 6d.

West on the Diseases of Women. Fourth Edition, revised by the Author, with numerous Additions by J. MATTHEWS DUNCAN, M.D., F.R.C.P., F.R.S.E., Obstetric Physician to St. Bartholomew's Hospital. 8vo, 16s.

Obstetric Aphorisms:
For the Use of Students commencing Midwifery Practice. By JOSEPH G. SWAYNE, M.D. Eighth Edition. Fcap. 8vo, with Engravings, 3s. 6d.

Handbook of Midwifery for Midwives: By J. E. BURTON, L.R.C.P. Lond., Surgeon to the Hospital for Women, Liverpool. Second Edition. With Engravings. Fcap. 8vo, 6s.

A Handbook of Uterine Therapeutics, and of Diseases of Women. By E. J. TILT, M.D., M.R.C.P. Fourth Edition. Post 8vo, 10s.

By the same Author.

The Change of Life
In Health and Disease: A Clinical Treatise on the Diseases of the Nervous System incidental to Women at the Decline of Life. Fourth Edition. 8vo, 10s. 6d.

Diseases of the Uterus, Ovaries, and Fallopian Tubes: A Practical Treatise by A. COURTY, Professor of Clinical Surgery, Montpellier. Translated from Third Edition by his Pupil, AGNES McLAREN, M.D., M.K.Q.C.P.I., with Preface by J. MATTHEWS DUNCAN, M.D., F.R.C.P. 8vo, with 424 Engravings, 24s.

The Female Pelvic Organs:
Their Surgery, Surgical Pathology, and Surgical Anatomy. In a Series of Coloured Plates taken from Nature; with Commentaries, Notes, and Cases. By HENRY SAVAGE, M.D., F.R.C.S., Consulting Officer of the Samaritan Free Hospital. Fifth Edition. Roy. 4to, with 17 Lithographic Plates (15 coloured) and 52 Woodcuts, £1 15s.

A Practical Treatise on the Diseases of Women. By T. GAILLARD THOMAS, M.D., Professor of Diseases of Women in the College of Physicians and Surgeons, New York. Fifth Edition. Roy. 8vo, with 266 Engravings, 25s.

Diseases and Accidents
Incident to Women, and the Practice of Medicine and Surgery applied to them. By W. H. BYFORD, A.M., M.D., Professor of Gynaecology in Rush Medical College, and HENRY T. BYFORD, M.D., Surgeon to the Woman's Hospital, Chicago. Fourth Edition. 8vo, with 306 Engravings, 25s.

Gynæcological Operations:
(Handbook of). By ALBAN H. G. DORAN, F.R.C.S., Surgeon to the Samaritan Hospital. 8vo, with 167 Engravings, 15s.

Abdominal Surgery.
By J. GREIG SMITH, M.A., F.R.S.E., Surgeon to the Bristol Royal Infirmary. Second Edition. 8vo, with 79 Engravings, 21s.

The Student's Guide to Diseases of Children. By JAS. F. GOODHART, M.D., F.R.C.P., Physician to Guy's Hospital, and to the Evelina Hospital for Sick Children. Second Edition. Fcap. 8vo, 10s. 6d.

Diseases of Children.
For Practitioners and Students. By W. H. DAY, M.D., Physician to the Samaritan Hospital. Second Edition. Crown 8vo, 12s. 6d.

A Practical Treatise on Disease in Children. By EUSTACE SMITH, M.D., Physician to the King of the Belgians, Physician to the East London Hospital for Children. 8vo, 22s.

By the same Author.

Clinical Studies of Disease in Children. Second Edition. Post 8vo, 7s. 6d.   *Also.*

The Wasting Diseases of Infants and Children. Fifth Edition. Post 8vo, 8s. 6d.

A Practical Manual of the Diseases of Children. With a Formulary. By EDWARD ELLIS, M.D. Fifth Edition. Crown 8vo, 10s.

A Manual for Hospital Nurses and others engaged in Attending the Sick, and a Glossary. By EDWARD J. DOMVILLE, Surgeon to the Exeter Lying-in Charity. Sixth Edition. Cr. 8vo, 2s. 6d.

A Manual of Nursing, Medical and Surgical. By CHARLES J. CULLINGWORTH, M.D., Obstetric Physician to St. Thomas's Hospital. Second Edition. Fcap. 8vo, with Engravings, 3s. 6d.

By the same Author.

A Short Manual for Monthly Nurses. Second Edition. Fcap. 8vo, 1s. 6d.

Diseases and their Commencement.

Lectures to Trained Nurses. By DONALD W. C. HOOD, M.D., M.R.C.P., Physician to the West London Hospital. Crown 8vo, 2s. 6d.

Outlines of Infectious Diseases :

For the use of Clinical Students. By J. W. ALLAN, M.B., Physician Superintendent Glasgow Fever Hospital. Fcap. 8vo., 3s.

Hospital Sisters and their Duties.

By EVA C. E. LÜCKES, Matron to the London Hospital. Second Edition. Crown 8vo, 2s. 6d.

Infant Feeding and its Influence on Life ;

By C. H. F. ROUTH, M.D., Physician to the Samaritan Hospital. Fourth Edition. Fcap. 8vo. [*Preparing.*

Manual of Botany :

Including the Structure, Classification, Properties, Uses, and Functions of Plants. By ROBERT BENTLEY, Professor of Botany in King's College and to the Pharmaceutical Society. Fifth Edition. Crown 8vo, with 1,178 Engravings, 15s.

By the same Author.

The Student's Guide to Structural, Morphological, and Physiological Botany.

With 660 Engravings. Fcap. 8vo, 7s. 6d.

Also.

The Student's Guide to Systematic Botany,

including the Classification of Plants and Descriptive Botany. Fcap. 8vo, with 350 Engravings, 3s. 6d.

Medicinal Plants :

Being descriptions, with original figures, of the Principal Plants employed in Medicine, and an account of their Properties and Uses. By Prof. BENTLEY and Dr. H. TRIMEN, F.R.S. In 4 vols., large 8vo, with 306 Coloured Plates, bound in Half Morocco, Gilt Edges, £11 11s.

Royle's Manual of Materia Medica and Therapeutics.

Sixth Edition, including additions and alterations in the B.P. 1885. By JOHN HARLEY, M.D., Physician to St. Thomas's Hospital. Crown 8vo, with 139 Engravings, 15s.

Materia Medica and Therapeutics :

Vegetable Kingdom — Organic Compounds — Animal Kingdom. By CHARLES D. F. PHILLIPS, M.D., F.R.S. Edin., late Lecturer on Materia Medica and Therapeutics at the Westminster Hospital Medical School. 8vo, 25s.

The Student's Guide to Materia Medica and Therapeutics.

By JOHN C. THOROWGOOD, M.D., F.R.C.P. Second Edition. Fcap. 8vo, 7s.

Materia Medica.

A Manual for the use of Students. By ISAMBARD OWEN, M.D., F.R.C.P., Lecturer on Materia Medica, &c., to St. George's Hospital. Second Edition. Crown 8vo, 6s. 6d.

A Companion to the British Pharmacopœia.

By PETER SQUIRE, Revised by his Sons, P. W. and A. H. SQUIRE. 14th Edition. 8vo, 10s. 6d.

By the same Authors.

The Pharmacopœias of the London Hospitals,

arranged in Groups for Easy Reference and Comparison. Fifth Edition. 18mo, 6s.

The Prescriber's Pharmacopœia :

The Medicines arranged in Classes according to their Action, with their Composition and Doses. By NESTOR J. C. TIRARD, M.D., F.R.C.P., Professor of Materia Medica and Therapeutics in King's College, London. Sixth Edition. 32mo, bound in leather, 3s.

A Treatise on the Principles and Practice of Medicine.

Sixth Edition. By AUSTIN FLINT, M.D., W.H. WELCH, M.D., and AUSTIN FLINT, jun., M.D. 8vo, with Engravings, 26s.

Climate and Fevers of India,

with a series of Cases (Croonian Lectures, 1882). By Sir JOSEPH FAYRER, K.C.S.I., M.D. 8vo, with 17 Temperature Charts, 12s.

By the same Author.

The Natural History and Epidemiology of Cholera :

Being the Annual Oration of the Medical Society of London, 1888. 8vo, 3s. 6d.

Family Medicine for India.

A Manual. By WILLIAM J. MOORE, M.D., C.I.E., Honorary Surgeon to the Viceroy of India. Published under the Authority of the Government of India. Fifth Edition. Post 8vo, with Engravings. [*In the Press.*

By the same Author.

A Manual of the Diseases of India :

With a Compendium of Diseases generally. Second Edition. Post 8vo, 10s.

Practical Therapeutics :

A Manual. By EDWARD J. WARING, C.I.E., M.D., F.R.C.P., and DUDLEY W. BUXTON, M.D., B.S. Lond. Fourth Edition. Crown 8vo, 14s.

By the same Author.

Bazaar Medicines of India,

And Common Medical Plants : With Full Index of Diseases, indicating their Treatment by these and other Agents procurable throughout India, &c. Fourth Edition. Fcap. 8vo, 5s.

A Commentary on the Diseases
of India. By NORMAN CHEVERS,
C.I.E., M.D., F.R.C.S., Deputy Sur-
geon-General H.M. Indian Army. 8vo,
24s.

The Principles and Practice of
Medicine. By C. HILTON FAGGE,
M.D. Second Edition. Edited by P. H.
PYE-SMITH, M.D., F.R.C.P., Physician
to, and Lecturer on Medicine at, Guy's
Hospital. 2 vols. 8vo. Cloth, 38s. ;
Half Leather, 44s.

The Student's Guide to the
Practice of Medicine. By M.
CHARTERIS, M.D., Professor of Therapeutics and Materia Medica in the University of Glasgow. Fourth Edition.
Fcap. 8vo, with Engravings on Copper
and Wood, 9s.

Hooper's Physicians' Vade-
Mecum. A Manual of the Principles
and Practice of Physic. Tenth Edition.
By W. A. GUY, F.R.C.P., F.R.S., and
J. HARLEY, M.D., F.R.C.P. With 118
Engravings. Fcap. 8vo, 12s. 6d.

Preventive Medicine.
Collected Essays. By WILLIAM SQUIRE,
M.D., F.R.C.P., Physician to St.
George, Hanover-square, Dispensary.
8vo, 6s, 6d.

The Student's Guide to Clinical
Medicine and Case-Taking. By
FRANCIS WARNER, M.D., F.R.C.P.,
Physician to the London Hospital.
Second Edition. Fcap. 8vo, 5s.

The Student's Guide to Dis-
eases of the Chest. By VINCENT
D. HARRIS, M.D. Lond., F.R.C.P.,
Physician to the City of London Hospital
for Diseases of the Chest, Victoria Park.
Fcap. 8vo, with 55 Illustrations (some
Coloured), 7s. 6d.

How to Examine the Chest :
A Practical Guide for the use of Students.
By SAMUEL WEST, M.D., F.R.C.P.,
Physician to the City of London
Hospital for Diseases of the Chest;
Assistant Physician to St. Bartholomew's
Hospital. With 42 Engravings. Fcap.
8vo, 5s.

Contributions to Clinical and
Practical Medicine. By A. T.
HOUGHTON WATERS, M.D., Physician
to the Liverpool Royal Infirmary.
8vo, with Engravings, 7s.

Fever : A Clinical Study.
By T. J. MACLAGAN, M.D. 8vo, 7s. 6d.

The Student's Guide to Medical
Diagnosis. By SAMUEL FENWICK,
M.D., F.R.C.P., Physician to the Lon-
don Hospital, and BEDFORD FENWICK,
M.D., M.R.C.P. Sixth Edition. Fcap.
8vo, with 114 Engravings, 7s.

By the same Author.

The Student's Outlines of Medical Treatment. Second Edition.
Fcap. 8vo, 7s.

Also.

On Chronic Atrophy of the
Stomach, and on the Nervous Affections
of the Digestive Organs. 8vo, 8s.

Also.

The Saliva as a Test for
Functional Diseases of the Liver.
Crown 8vo, 2s.

The Microscope in Medicine.
By LIONEL S. BEALE, M.B., F.R.S.,
Physician to King's College Hospital.
Fourth Edition. 8vo, with 86 Plates, 21s.

Also.

On Slight Ailments :
Their Nature and Treatment. Second
Edition. 8vo, 5s.

Medical Lectures and Essays.
By G. JOHNSON, M.D., F.R.C.P., F.R.S.,
Consulting Physician to King's College
Hospital. 8vo, with 46 Engravings, 25s.

Notes on Asthma :
Its Forms and Treatment. By JOHN C.
THOROWGOOD, M.D., Physician to the
Hospital for Diseases of the Chest. Third
Edition. Crown 8vo, 4s. 6d.

Winter Cough
(Catarrh, Bronchitis, Emphysema, Asthma). By HORACE DOBELL, M.D.,
Consulting Physician to the Royal Hospital for Diseases of the Chest. Third Edition. 8vo, with Coloured Plates, 10s. 6d.

By the same Author.

Loss of Weight, Blood-Spitting,
and Lung Disease. Second Edition.
8vo, with Chromo-lithograph, 10s. 6d.

Also.

The Mont Dore Cure, and the
Proper Way to Use it. 8vo, 7s. 6d.

Vaccinia and Variola :
A Study of their Life History. By JOHN
B. BUIST, M.D., F.R.S.E., Teacher of
Vaccination for the Local Government
Board. Crown 8vo, with 24 Coloured
Plates, 7s. 6d.

Treatment of Some of the Forms
of Valvular Disease of the Heart.
By A. E. SANSOM, M.D., F.R.C.P.,
Physician to the London Hospital.
Second Edition. Fcap. 8vo, with 26
Engravings, 4s. 6d.

Diseases of the Heart and Aorta:
Clinical Lectures. By G. W. BALFOUR, M.D., F.R.C.P., F.R.S. Edin., late Senior Physician and Lecturer on Clinical Medicine, Royal Infirmary, Edinburgh. Second Edition. 8vo, with Chromo-lithograph and Wood Engravings, 12s. 6d.

Medical Ophthalmoscopy :
A Manual and Atlas. By W. R. GOWERS, M.D., F.R.C.P., F.R.S., Professor of Clinical Medicine in University College, Physician to University College Hospital and to the National Hospital for the Paralyzed and Epileptic. Second Edition, with Coloured Plates and Woodcuts. 8vo, 18s.

By the same Author.

Diagnosis of Diseases of the Brain.
Second Edition. 8vo, with Engravings, 7s. 6d.

Also.

Diagnosis of Diseases of the Spinal Cord.
Third Edition. 8vo, with Engravings, 4s. 6d.

Also.

A Manual of Diseases of the Nervous System.
Vol. I. Diseases of the Spinal Cord and Nerves. Roy. 8vo, with 171 Engravings (many figures), 12s. 6d.
Vol. II. Diseases of the Brain and Cranial Nerves : General and Functional Diseases of the Nervous System. 8vo, with 170 Engravings, 17s. 6d.

Diseases of the Nervous System.
Lectures delivered at Guy's Hospital. By SAMUEL WILKS, M.D., F.R.S. Second Edition. 8vo, 18s.

Nerve Vibration and Excitation,
as Agents in the Treatment of Functional Disorder and Organic Disease. By J. MORTIMER GRANVILLE, M.D. 8vo, 5s.

By the same Author.

Gout in its Clinical Aspects.
Crown 8vo, 6s.

Regimen to be adopted in Cases
of Gout. By WILHELM EBSTEIN, M.D., Professor of Clinical Medicine in Göttingen. Translated by JOHN SCOTT, M.A., M.B. 8vo, 2s. 6d.

Diseases of the Nervous System.
Clinical Lectures. By THOMAS BUZZARD, M.D., F.R.C.P., Physician to the National Hospital for the Paralysed and Epileptic. With Engravings, 8vo. 15s.

By the same Author.

Some Forms of Paralysis from
Peripheral Neuritis : of Gouty, Alcoholic, Diphtheritic, and other origin. Crown 8vo, 5s.

Diseases of the Liver :
With and without Jaundice. By GEORGE HARLEY, M.D., F.R.C.P., F.R.S. 8vo, with 2 Plates and 36 Engravings, 21s.

By the same Author.

Inflammations of the Liver, and
their Sequelæ. Crown 8vo, with Engravings, 5s.

Gout, Rheumatism,
And the Allied Affections ; with Chapters on Longevity and Sleep. By PETER HOOD, M.D. Third Edition. Crown 8vo, 7s. 6d.

Diseases of the Stomach :
The Varieties of Dyspepsia, their Diagnosis and Treatment. By S. O. HABERSHON, M.D., F.R.C.P. Third Edition. Crown 8vo, 5s.

By the same Author.

Pathology of the Pneumogastric Nerve :
Lumleian Lectures for 1876. Second Edition. Post 8vo, 4s.

Also.

Diseases of the Abdomen,
Comprising those of the Stomach and other parts of the Alimentary Canal, Œsophagus, Cæcum, Intestines, and Peritoneum. Fourth Edition. 8vo, with 5 Plates, 21s.

Also.

Diseases of the Liver,
Their Pathology and Treatment. Lettsomian Lectures. Second Edition. Post 8vo, 4s.

On the Relief of Excessive and
Dangerous Tympanites by Puncture of the Abdomen. By JOHN W. OGLE, M.A., M.D., F.R.C.P., Consulting Physician to St. George's Hospital. 8vo, 5s. 6d.

Acute Intestinal Strangulation,
And Chronic Intestinal Obstruction (Mode of Death from). By THOMAS BRYANT, F.R.C.S., Senior Surgeon to Guy's Hospital. 8vo, 3s.

A Treatise on the Diseases of
the Nervous System. By JAMES ROSS, M.D., F.R.C.P., Assistant Physician to the Manchester Royal Infirmary. Second Edition. 2 vols. 8vo, with Lithographs, Photographs, and 332 Woodcuts, 52s. 6d.

By the same Author.

Handbook of the Diseases of
the Nervous System. Roy. 8vo, with 184 Engravings, 18s.

Also.

Aphasia :
Being a Contribution to the Subject of the Dissolution of Speech from Cerebral Disease. 8vo, with Engravings, 4s. 6d.

Spasm in Chronic Nerve Disease.
(Gulstonian Lectures.) By SEYMOUR J. SHARKEY, M.A., M.B., F.R.C.P., Assistant Physician to, and Joint Lecturer on Pathology at, St. Thomas's Hospital. 8vo, with Engravings, 5s.

Food and Dietetics,
Physiologically and Therapeutically Considered. By F. W. PAVY, M.D., F.R.S., Physician to Guy's Hospital. Second Edition. 8vo, 15s.

By the same Author.

Croonian Lectures on Certain Points connected with Diabetes.
8vo, 4s. 6d.

Headaches :
Their Nature, Causes, and Treatment. By W. H. DAY, M.D., Physician to the Samaritan Hospital. Fourth Edition. Crown 8vo, with Engravings, 7s. 6d.

Health Resorts at Home and Abroad.
By M. CHARTERIS, M.D., Professor of Therapeutics and Materia Medica in the University of Glasgow. Second Edition. Crown 8vo, with Map, 5s. 6d.

Winter and Spring
On the Shores of the Mediterranean. By HENRY BENNET, M.D. Fifth Edition. Post 8vo, with numerous Plates, Maps, and Engravings, 12s. 6d.

Medical Guide to the Mineral Waters of France and its Wintering Stations.
With a Special Map. By A. VINTRAS, M.D., Physician to the French Embassy, and to the French Hospital, London. Crown 8vo, 8s.

The Ocean as a Health-Resort :
A Practical Handbook of the Sea, for the use of Tourists and Health-Seekers. By WILLIAM S. WILSON, L.R.C.P. Second Edition, with Chart of Ocean Routes, &c. Crown 8vo, 7s. 6d.

Ambulance Handbook for Volunteers and Others.
By J. ARDAVON RAYE, L.K. & Q.C.P.I., L.R.C.S.I., late Surgeon to H.B.M. Transport No. 14, Zulu Campaign, and Surgeon E.I.R. Rifles. 8vo, with 16 Plates (50 figures), 3s. 6d.

Ambulance Lectures:
To which is added a NURSING LECTURE. By JOHN M. H. MARTIN, Honorary Surgeon to the Blackburn Infirmary. Second Edition. Crown 8vo, with 59 Engravings, 2s.

Commoner Diseases and Accidents to Life and Limb: their Prevention and Immediate Treatment.
By M. M. BASIL, M.A., M.B., C.M. Crown 8vo, 2s. 6d.

How to Use a Galvanic Battery in Medicine and Surgery.
By HERBERT TIBBITS, M.D., F.R.C.P.E., Senior Physician to the West London Hospital for Paralysis and Epilepsy. Third Edition. 8vo, with Engravings, 4s.

By the same Author.

A Map of Ziemssen's Motor Points of the Human Body : A
Guide to Localised Electrisation. Mounted on Rollers, 35 × 21. With 20 Illustrations, 5s.   *Also.*

Electrical and Anatomical Demonstrations.
A Handbook for Trained Nurses and Masseuses. Crown 8vo, with 44 Illustrations, 5s.

Surgical Emergencies :
Together with the Emergencies attendant on Parturition and the Treatment of Poisoning. By W. PAUL SWAIN, F.R.C.S., Surgeon to the South Devon and East Cornwall Hospital. Fourth Edition. Crown 8vo, with 120 Engravings, 5s.

Operative Surgery in the Calcutta Medical College Hospital.
Statistics, Cases, and Comments. By KENNETH McLEOD, A.M., M.D., F.R.C.S.E., Surgeon-Major, Indian Medical Service, Professor of Surgery in Calcutta Medical College. 8vo, with Illustrations, 12s. 6d.

Surgical Pathology and Morbid Anatomy (Student's Guide).
By ANTHONY A. BOWLBY, F.R.C.S., Surgical Registrar and Demonstrator of Surgical Pathology to St. Bartholomew's Hospital. Fcap. 8vo, with 135 Engravings, 9s.

A Course of Operative Surgery.
By CHRISTOPHER HEATH, Surgeon to University College Hospital. Second Edition. With 20 coloured Plates (180 figures) from Nature, by M. LÉVEILLÉ, and several Woodcuts. Large 8vo, 30s.

By the same Author.

The Student's Guide to Surgical Diagnosis.
Second Edition. Fcap. 8vo, 6s. 6d.   *Also.*

Manual of Minor Surgery and Bandaging.
For the use of House-Surgeons, Dressers, and Junior Practitioners. Eighth Edition. Fcap. 8vo, with 142 Engravings, 6s.

Also.

Injuries and Diseases of the Jaws.
Third Edition. 8vo, with Plate and 206 Wood Engravings, 14s.

Also,

Lectures on Certain Diseases of the Jaws.
Delivered at the R.C.S., Eng., 1887. 8vo, with 64 Engravings, 2s. 6d.

The Practice of Surgery:

A Manual. By THOMAS BRYANT, Surgeon to Guy's Hospital. Fourth Edition. 2 vols. crown 8vo, with 750 Engravings (many being coloured), and including 6 chromo plates, 32s.

Surgery: its Theory and Practice (Student's Guide).

By WILLIAM J. WALSHAM, F.R.C.S., Assistant Surgeon to St. Bartholomew's Hospital. Fcap. 8vo, with 236 Engravings, 10s. 6d.

The Surgeon's Vade-Mecum:

A Manual of Modern Surgery. By R. DRUITT, F.R.C.S. Twelfth Edition. By STANLEY BOYD, M.B., F.R.C.S. Assistant Surgeon and Pathologist to Charing Cross Hospital. Crown 8vo, with 373 Engravings 16s.

Regional Surgery:

Including Surgical Diagnosis. A Manual for the use of Students. By F. A. SOUTHAM, M.A., M.B., F.R.C.S., Assistant Surgeon to the Manchester Royal Infirmary. Part I. The Head and Neck. Crown 8vo, 6s. 6d. — Part II. The Upper Extremity and Thorax. Crown 8vo, 7s. 6d. Part III. The Abdomen and Lower Extremity. Crown 8vo, 7s.

Illustrations of Clinical Surgery.

By JONATHAN HUTCHINSON, F.R.S., Senior Surgeon to the London Hospital. In fasciculi. 6s. 6d each. Fasc. I. to X. bound, with Appendix and Index, £3 10s. Fasc. XI. to XXIII. bound, with Index, £4 10s.

A Treatise on Dislocations.

By LEWIS A. STIMSON, M.D., Professor of Clinical Surgery in the University of the City of New York. Roy. 8vo, with 163 Engravings, 15s.

By the same Author.

A Treatise on Fractures.

Roy. 8vo, with 360 Engravings, 21s.

Lectures on Orthopædic Surgery.

By BERNARD E. BRODHURST, F.R.C.S., Surgeon to the Royal Orthopædic Hospital. Second Edition. 8vo, with Engravings, 12s. 6d.

By the same Author.

On Anchylosis, and the Treatment for the Removal of Deformity and the Restoration of Mobility in Various Joints.

Fourth Edition. 8vo, with Engravings, 5s.

Also.

Curvatures and Disease of the Spine.

Fourth Edition. 8vo, with Engravings, 7s. 6d.

Diseases of Bones and Joints.

By CHARLES MACNAMARA, F.R.C.S., Surgeon to, and Lecturer on Surgery at, the Westminster Hospital. 8vo, with Plates and Engravings, 12s.

Injuries of the Spine and Spinal Cord, and NERVOUS SHOCK,

in their Surgical and Medico-Legal Aspects. By HERBERT W. PAGE, M.C. Cantab., F.R.C.S., Surgeon to St. Mary's Hospital. Second Edition, post 8vo, 10s.

Spina Bifida:

Its Treatment by a New Method. By JAS. MORTON, M.D., L.R.C.S.E., Professor of Materia Medica in Anderson's College, Glasgow. 8vo, with Plates, 7s. 6d.

Face and Foot Deformities.

By FREDERICK CHURCHILL, C.M., Surgeon to the Victoria Hospital for Children. 8vo, with Plates and Illustrations, 10s. 6d.

Clubfoot:

Its Causes, Pathology, and Treatment. By WM. ADAMS, F.R.C.S., Surgeon to the Great Northern Hospital. Second Edition. 8vo, with 106 Engravings and 6 Lithographic Plates, 15s.

By the same Author.

On Contraction of the Fingers,

and its Treatment by Subcutaneous Operation; and on Obliteration of Depressed Cicatrices, by the same Method. 8vo, with 30 Engravings, 4s. 6d.

Also.

Lateral and other Forms of Curvature of the Spine:

Their Pathology and Treatment. Second Edition. 8vo, with 5 Lithographic Plates and 72 Wood Engravings, 10s. 6d.

Electricity and its Manner of Working in the Treatment of Disease.

By WM. E. STEAVENSON, M.D., Physician and Electrician to St. Bartholomew's Hospital. 8vo, 4s. 6d.

On Diseases and Injuries of the Eye:

A Course of Systematic and Clinical Lectures to Students and Medical Practitioners. By J. R. WOLFE, M.D., F.R.C.S.E., Lecturer on Ophthalmic Medicine and Surgery in Anderson's College, Glasgow. With 10 Coloured Plates and 157 Wood Engravings. 8vo, £1 1s.

Hints on Ophthalmic Out-Patient Practice.

By CHARLES HIGGENS, Ophthalmic Surgeon to Guy's Hospital. Third Edition. Fcap. 8vo, 3s.

The Student's Guide to Diseases

of the Eye. By EDWARD NETTLESHIP, F.R.C.S., Ophthalmic Surgeon to St. Thomas's Hospital. Fourth Edition. Fcap. 8vo, with 164 Engravings and a Set of Coloured Papers illustrating Colour-Blindness, 7s. 6d.

Manual of the Diseases of the

Eye. By CHARLES MACNAMARA, F.R.C.S., Surgeon to Westminster Hospital. Fourth Edition. Crown 8vo, with 4 Coloured Plates and 66 Engravings, 10s. 6d.

Normal and Pathological Histology of the Human Eye and

Eyelids. By C. FRED. POLLOCK, M.D., F.R.C.S. and F.R.S.E., Surgeon for Diseases of the Eye to Anderson's College Dispensary, Glasgow. Crown 8vo, with 100 Plates (230 drawings), 15s.

Atlas of Ophthalmoscopy.

Composed of 12 Chromo-lithographic Plates (59 Figures drawn from nature) and Explanatory Text. By RICHARD LIEBREICH, M.R.C.S. Translated by H. ROSBOROUGH SWANZY, M.B. Third edition, 4to, 40s.

Refraction of the Eye:

A Manual for Students. By GUSTAVUS HARTRIDGE, F.R.C.S., Assistant Surgeon to the Royal Westminster Ophthalmic Hospital. Third Edition. Crown 8vo, with 96 Illustrations, Test-types, &c., 5s. 6d.

Squint:

(Clinical Investigations on). By C. SCHWEIGGER, M.D., Professor of Ophthalmology in the University of Berlin. Edited by GUSTAVUS HARTRIDGE, F.R.C.S. 8vo, 5s.

Practitioner's Handbook of Diseases of the Ear and Naso-

Pharynx. By H. MACNAUGHTON JONES, M.D., late Professor of the Queen's University in Ireland, Surgeon to the Cork Ophthalmic and Aural Hospital. Third Edition of "Aural Surgery." Roy. 8vo, with 128 Engravings, 6s.

By the same Author.

Atlas of Diseases of the Membrana

Tympani. In Coloured Plates, containing 62 Figures, with Text. Crown 4to, 21s.

Endemic Goitre or Thyreocele:

Its Etiology, Clinical Characters, Pathology, Distribution, Relations to Cretinism, Myxœdema, &c., and Treatment. By WILLIAM ROBINSON, M.D. 8vo, 5s.

Diseases and Injuries of the

Ear. By Sir WILLIAM B. DALBY, Aural Surgeon to St. George's Hospital. Third Edition. Crown 8vo, with Engravings, 7s. 6d.

By the Same Author.

Short Contributions to Aural

Surgery, between 1875 and 1886. 8vo, with Engravings, 3s. 6d.

Diseases of the Throat and

Nose: A Manual. By Sir MORELL MACKENZIE, M.D., Senior Physician to the Hospital for Diseases of the Throat. Vol. II. Diseases of the Nose and Naso-Pharynx; with a Section on Diseases of the Œsophagus. Post 8vo, with 93 Engravings, 12s. 6d.

By the same Author.

Diphtheria:

Its Nature and Treatment, Varieties, and Local Expressions. 8vo, 5s.

Sore Throat:

Its Nature, Varieties, and Treatment. By PROSSER JAMES, M.D., Physician to the Hospital for Diseases of the Throat. Fifth Edition. Post 8vo, with Coloured Plates and Engravings, 6s. 6d.

Studies in Pathological Anatomy,

Especially in Relation to Laryngeal Neoplasms. By R. NORRIS WOLFENDEN, M.D., Senior Physician to the Throat Hospital, and SIDNEY MARTIN, M.D., Pathologist to the City of London Hospital, Victoria Park. I. Papilloma. Roy. 8vo, with Coloured Plates, 2s. 6d.

A System of Dental Surgery.

By Sir JOHN TOMES, F.R.S., and C. S. TOMES, M.A., F.R.S. Third Edition. Crown 8vo, with 292 Engravings, 15s.

Dental Anatomy, Human and

Comparative: A Manual. By CHARLES S. TOMES, M.A., F.R.S. Second Edition. Crown 8vo, with 191 Engravings, 12s. 6d.

The Student's Guide to Dental

Anatomy and Surgery. By HENRY SEWILL, M.R.C.S., L.D.S. Second Edition. Fcap. 8vo, with 78 Engravings, 5s. 6d.

A Manual of Nitrous Oxide

Anæsthesia, for the use of Students and General Practitioners. By J. FREDERICK W. SILK, M.D. Lond., M.R.C.S., Anæsthetist to the Great Northern Central Hospital, and to the National Dental Hospital. 8vo, with 26 Engravings, 5s.

Mechanical Dentistry in Gold

and Vulcanite. By F. H. BALKWILL, L.D.S.R.C.S. 8vo, with 2 Lithographic Plates and 57 Engravings, 10s.

Principles and Practice of Dentistry : including Anatomy, Physiology, Pathology, Therapeutics, Dental Surgery, and Mechanism. By C. A. HARRIS, M.D., D.D.S. Edited by F. J. S. GORGAS, A.M., M.D., D.D.S., Professor in the Dental Department of Maryland University. Eleventh Edition. 8vo, with 750 Illustrations, 31s. 6d.

A Practical Treatise on Mechanical Dentistry. By JOSEPH RICHARDSON, M.D., D.D.S., late Emeritus Professor of Prosthetic Dentistry in the Indiana Medical College. Fourth Edition. Roy. 8vo, with 458 Engravings, 21s.

Elements of Dental Materia Medica and Therapeutics, with Pharmacopœia. By JAMES STOCKEN, L.D.S.R.C.S., Pereira Prizeman for Materia Medica, and THOMAS GADDES, L.D.S. Eng. and Edin. Third Edition. Fcap. 8vo, 7s. 6d.

Atlas of Skin Diseases.
By TILBURY FOX, M.D., F.R.C.P. With 72 Coloured Plates. Royal 4to, half morocco, £6 6s.

Diseases of the Skin :
With an Analysis of 8,000 Consecutive Cases and a Formulary. By L. D. BULKLEY, M.D., Physician for Skin Diseases at the New York Hospital. Crown 8vo, 6s. 6d.

By the same Author.

Acne : its Etiology, Pathology, and Treatment : Based upon a Study of 1,500 Cases. 8vo, with Engravings, 10s.

On Certain Rare Diseases of the Skin. By JONATHAN HUTCHINSON, F.R.S., Senior Surgeon to the London Hospital, and to the Hospital for Diseases of the Skin. 8vo, 10s. 6d.

Diseases of the Skin :
A Practical Treatise for the Use of Students and Practitioners. By J. N. HYDE, A.M., M.D., Professor of Skin and Venereal Diseases, Rush Medical College, Chicago. Second Edition. 8vo, with 2 Coloured Plates and 96 Engravings, 20s.

Parasites :
A Treatise on the Entozoa of Man and Animals, including some Account of the Ectozoa. By T. SPENCER COBBOLD, M.D., F.R.S. 8vo, with 85 Engravings, 15s.

Manual of Animal Vaccination, preceded by Considerations on Vaccination in general. By E. WARLOMONT, M.D., Founder of the State Vaccine Institute of Belgium. Translated and edited by ARTHUR J. HARRIES, M.D. Crown 8vo, 4s. 6d.

Leprosy in British Guiana.
By JOHN D. HILLIS, F.R.C.S., M.R.I.A., Medical Superintendent of the Leper Asylum, British Guiana. Imp. 8vo, with 22 Lithographic Coloured Plates and Wood Engravings, £1 11s. 6d.

Cancer of the Breast.
By THOMAS W. NUNN, F.R.C.S., Consulting Surgeon to the Middlesex Hospital. 4to, with 21 Coloured Plates, £2 2s.

On Cancer :
Its Allies, and other Tumours; their Medical and Surgical Treatment. By F. A. PURCELL, M.D., M.C., Surgeon to the Cancer Hospital, Brompton. 8vo, with 21 Engravings, 10s. 6d.

Sarcoma and Carcinoma :
Their Pathology, Diagnosis, and Treatment. By HENRY T. BUTLIN, F.R.C.S., Assistant Surgeon to St. Bartholomew's Hospital. 8vo, with 4 Plates, 8s.

By the same Author.

Malignant Disease of the Larynx (Sarcoma and Carcinoma). 8vo, with 5 Engravings, 5s.

Also.

Operative Surgery of Malignant Disease. 8vo, 14s.

Cancerous Affections of the Skin.
(Epithelioma and Rodent Ulcer.) By GEORGE THIN, M.D. Post 8vo, with 8 Engravings, 5s.

By the same Author.

Pathology and Treatment of Ringworm. 8vo, with 21 Engravings, 5s.

Cancer of the Mouth, Tongue, and Alimentary Tract : their Pathology, Symptoms, Diagnosis. and Treatment. By FREDERIC B. JESSETT, F.R.C.S., Surgeon to the Cancer Hospital, Brompton. 8vo, 10s.

Lectures on the Surgical Disorders of the Urinary Organs. By REGINALD HARRISON, F.R.C.S., Surgeon to the Liverpool Royal Infirmary. Third Edition, with 117 Engravings. 8vo, 12s. 6d.

Hydrocele :
Its several Varieties and their Treatment. By SAMUEL OSBORN, late Surgical Registrar to St. Thomas's Hospital. Fcap. 8vo, with Engravings, 3s.

By the same Author.

Diseases of the Testis.
Fcap. 8vo, with Engravings, 3s. 6d.

Diseases of the Urinary Organs.

Clinical Lectures. By Sir HENRY THOMPSON, F.R.C.S., Emeritus Professor of Clinical Surgery in University College. Seventh (Students') Edition. 8vo, with 84 Engravings, 2s. 6d.

By the same Author.

Diseases of the Prostate :

Their Pathology and Treatment. Sixth Edition. 8vo, with 39 Engravings, 6s.

Also.

Surgery of the Urinary Organs.

Some Important Points connected therewith. Lectures delivered in the R.C.S. 8vo, with 44 Engravings. Students' Edition, 2s. 6d.

Also.

Practical Lithotomy and Lithotrity;

or, An Inquiry into the Best Modes of Removing Stone from the Bladder. Third Edition. 8vo, with 87 Engravings, 10s.

Also.

The Preventive Treatment of

Calculous Disease, and the Use of Solvent Remedies. Third Edition. Crown 8vo, 2s. 6d.

Also.

Tumours of the Bladder :

Their Nature, Symptoms, and Surgical Treatment. 8vo, with numerous Illustrations, 5s.

Also.

Stricture of the Urethra, and Urinary Fistulæ :

their Pathology and Treatment. Fourth Edition. With 74 Engravings. 8vo, 6s.

Also.

The Suprapubic Operation of

Opening the Bladder for the Stone and for Tumours. 8vo, with 14 Engravings, 3s. 6d.

Electric Illumination of the Male

Bladder and Urethra, as a Means of Diagnosis of Obscure Vesico-Urethral Diseases. By E. HURRY FENWICK, F.R.C.S., Assistant Surgeon to the London Hospital and Surgeon to St. Peter's Hospital for Stone. 8vo, with 30 Engravings, 4s. 6d.

Modern Treatment of Stone in

the Bladder by Litholopaxy. By P. J. FREYER, M.A., M.D., M.Ch., Bengal Medical Service. 8vo, with Engravings, 5s.

The Surgical Diseases of the

Genito - Urinary Organs, including Syphilis. By E. L. KEYES, M.D., Professor of Genito-Urinary Surgery, Syphiology, and Dermatology in Bellevue Hospital Medical College, New York (a revision of VAN BUREN and KEYES' Text-book). Roy. 8vo, with 114 Engravings, 21s.

The Surgery of the Rectum.

By HENRY SMITH, Emeritus Professor of Surgery in King's College, Consulting Surgeon to the Hospital. Fifth Edition. 8vo, 6s.

Diseases of the Rectum and

Anus. By W. HARRISON CRIPPS, F.R.C.S., Assistant Surgeon to St. Bartholomew's Hospital, &c. 8vo, with 13 Lithographic Plates and numerous Wood Engravings, 12s. 6d.

Urinary and Renal Derangements and Calculous Disorders.

By LIONEL S. BEALE, F.R.C.P., F.R.S., Physician to King's College Hospital. 8vo, 5s.

The Diagnosis and Treatment

of Diseases of the Rectum. By WILLIAM ALLINGHAM, F.R.C.S., Surgeon to St. Mark's Hospital for Fistula. Fifth Edition. By HERBERT WM. ALLINGHAM, F.R.C.S., Surgeon to the Great Northern Central Hospital, Demonstrator of Anatomy at St. George's Hospital. 8vo, with 53 Engravings. 10s. 6d.

Syphilis and Pseudo-Syphilis.

By ALFRED COOPER, F.R.C.S., Surgeon to the Lock Hospital, to St. Mark's and the West London Hospitals. 8vo, 10s. 6d.

Diagnosis and Treatment of

Syphilis. By TOM ROBINSON, M.D., Physician to St. John's Hospital for Diseases of the Skin. Crown 8vo, 3s. 6d.

By the same Author.

Eczema : its Etiology, Pathology, and Treatment.

Crown 8vo, 3s. 6d.

Coulson on Diseases of the

Bladder and Prostate Gland. Sixth Edition. By WALTER J. COULSON, Surgeon to the Lock Hospital and to St. Peter's Hospital for Stone. 8vo, 16s.

The Medical Adviser in Life Assurance.

By Sir E. H. SIEVEKING, M.D., F.R.C.P. Second Edition. Crown 8vo, 6s.

A Medical Vocabulary :

An Explanation of all Terms and Phrases used in the various Departments of Medical Science and Practice, their Derivation, Meaning, Application, and Pronunciation. By R. G. MAYNE, M.D., LL.D. Sixth Edition. [*In the Press.*

A Dictionary of Medical Science:

Containing a concise Explanation of the various Subjects and Terms of Medicine, &c. By ROBLEY DUNGLISON, M.D., LL.D. Royal 8vo, 28s.

INDEX.

[Continued on the next page

The following CATALOGUES issued by J. & A. CHURCHILL will be forwarded post free on application :—

A. *J. & A. Churchill's General List of about 650 works on Anatomy, Physiology, Hygiene, Midwifery, Materia Medica, Medicine, Surgery, Chemistry, Botany, &c., &c., with a complete Index to their Subjects, for easy reference.* N.B.—*This List includes* B, C, & D.

B. *Selection from J. & A. Churchill's General List, comprising all recent Works published by them on the Art and Science of Medicine.*

C. *J. & A. Churchill's Catalogue of Text Books specially arranged for Students.*

D. *A selected and descriptive List of J. & A. Churchill's Works on Chemistry, Materia Medica, Pharmacy, Botany, Photography, Zoology, the Microscope, and other branches of Science.*

E. *The Medical Intelligencer, being a List of New Works and New Editions published by J. & A. Churchill.*

[Sent yearly to every Medical Practitioner in the United Kingdom whose name and address can be ascertained. A large number are also sent to the United States of America, Continental Europe, India, and the Colonies.]

AMERICA.—*J. & A. Churchill being in constant communication with various publishing houses in Boston, New York, and Philadelphia, are able, notwithstanding the absence of international copyright, to conduct negotiations favourable to English Authors.*

LONDON: 11, NEW BURLINGTON STREET.

Pardon & Sons, Printers,]   [*Wine Office Court, Fleet Street, E.C.*